This book is dedicated to all people working in manufacturing—past, present, and future.

ABOUT THE AUTHOR

Michael Quirk spent two years developing new courses and teaching as an associate professor of semiconductor manufacturing technology at Austin Community College, Austin, Texas. His education includes a B.A. in sociology from University of Arkansas, a B.S. in mechanical engineering from the University of Texas at Austin, and an M.S. in manufacturing systems engineering from Stanford University. He spent nine years with IBM working as an engineer and manager in manufacturing and process development positions, followed by five years working in manufacturing positions as a team member and leader for smaller firms in the United States and Europe. He is currently working in a small firm implementing manufacturing improvement in a team environment.

PREFACE

Manufacturing is the science of creating; with materials and a process, we create a product. Because people work *in* a manufacturing process, they naturally want to become a part of their creation.

Therein lies the problem. While we are an integral part of the manufacturing process, someone else is telling us how to do our work. This "someone" may not even understand the work of which we are a part. This produces emotional confusion. To survive, we separate our minds from the process, rationalizing that we are just like cogs on a wheel. Our work lacks commitment.

We overcome confusion by developing relationships with manufacturing colleagues. Our interaction occurs on two levels: the human and the technical. The human level addresses the contradiction of being a person amidst machines, while the technical level is a means of supporting one another to solve manufacturing problems. The organization to develop these relationships is the team. As a member of a functioning team, we contribute human and technical expertise for the common good of the process. We accept responsibility for our work.

Accepting responsibility for our actions moves the team to a new level, with a desire to know *why* the process functions as it does. The goal is to find the truth in manufacturing. We uncover this truth by constantly working to eliminate that which is false. By default, the process moves closer and closer toward perfection. In this manner, we go forward, giving a small part of ourselves each day through our work in the process. All this is done in the name of manufacturing improvement.

MQ

v

ACKNOWLEDGMENTS

The author wishes to acknowledge the contributions from the following people, without whose assistance this book could not have been written:

Hope Good, for her thorough manuscript proofing and editing.
Federico Miller, for his brainstorming sessions on statistics and other areas.
Julian Serda, for his input on the first chapter, plus overall review and feedback.
Don Studebaker, for his advice on improvement.
Venancio Ybarra, Jr., for his effort in all areas to put this book together.

Thanks to the people in the Semiconductor Manufacturing Technology Department who gave their encouragement and help along the way: Ed Ardizoni, Claudia Burnett, John Guyton, Dahlia Hernandez, Maurice Jones, Bynum Thomas, and Lucia Zavala.

Thanks to all the students at Austin Community College who provided feedback while taking the course, especially those students who shared a part of their work for the case studies. It has been a pleasure to work with you.

The author would like to acknowledge Pat McDonough, Weatherford College, Paul Molino, Mt. Hood Community College, and the summer of 1998 MATEC SPC class for their review of the manuscript and valuable input.

The author appreciates the effort and professionalism of Charles Stewart and Steve Robb at Prentice Hall and Kelli Jauron and Carey Lange at Carlisle Publishers Services.

A special thanks to Monique Quirk for her editing comments and her support throughout the year.

CONTENTS

1 INTRODUCTION TO MANUFACTURING   **2**

  1.1   What Is Manufacturing?   3
       1.1.1   High-Volume Manufacturing   4
  1.2   Competitive Manufacturing   9
  1.3   Elements of Manufacturing   14
       1.3.1   Manufacturing Variables   17
  1.4   Manufacturing and the Product   18
       1.4.1   High Technology—Semiconductors   18
       1.4.2   Semiconductor Process Technology   20
  1.5   Manufacturing and the Market   22

2 A SHORT HISTORY OF U.S. MANUFACTURING   **28**

  2.1   Pre–World War II   29
  2.2   1950s: Capacity Production   30
  2.3   1960s: Technology and Marketing   30
  2.4   1970s: Retreat   31
  2.5   1980s: Standing Up   32
  2.6   1990s: The Crossroads   32

3 MANUFACTURING IMPROVEMENT PROGRAMS   **36**

   3.1   Improvement from Japan   37
   3.2   Improvement from U.S. Firms   41
   3.3   Programs by Consultants   43
   3.4   Effectiveness of Improvement Programs   45

4 MANUFACTURING TEAMS   **48**

   4.1   Traditional Organization   49
   4.2   Teams   51
       4.2.1 Team Attitude   52
       4.2.2 Team Energy   52
   4.3   Types of Manufacturing Teams   53
   4.4   Team Members   57
   4.5   Effective Team Attributes   60
   4.6   Individual Skills for Effective Teams   63
   4.7   Leadership Skills for Teams   65
   4.8   Pitfalls of Teams   67

5 THE PROCESS   **76**

   5.1   Process Terminology   77
       5.1.1 Cycle Time Reduction: Series versus Parallel   89
       5.1.2 Cycle Time Reduction: Single-Piece
        Processing   90
   5.2   A Holistic Process View   99

6 SOURCES OF PROCESS WASTE   **106**

   6.1   Value versus Waste   107
       6.1.1 Identifying Waste   107
   6.2   Seven Sources of Waste   108

7 IMPROVEMENT   **124**

   7.1   Manufacturing Improvement   125
   7.2   Levels of Improvement   126
       7.2.1 Improvement Strategy   127
   7.3   Manufacturing Equipment   127
       7.3.1 Equipment Strategy   129

7.4   Equipment Performance   130
    7.4.1 Chronic and Sporadic Losses   132
7.5   Sources of Poor Equipment Performance   135
7.6   Overall Equipment Effectiveness   141
    7.6.1 Calculating and Using OEE   143

8 CONTINUAL IMPROVEMENT   150

8.1   Continual Improvement   151

9 BASIC STATISTICS FOR IMPROVEMENT   170

9.1   What Is Statistics?   171
9.2   Statistical Data   172
    9.2.1 Variables versus Attribute Data   173
9.3   Variation   174
    9.3.1 Common Cause Variation   174
    9.3.2 Special Cause Variation   175
    9.3.3 Process Stability   175
    9.3.4 Natural Process Variation   175
9.4   Data Collection   178
    9.4.1 Manufacturing Data   178
    9.4.2 Collecting Good Data   180
9.5   Data Analysis: Histogram   181
9.6   Normal Curve   184
9.7   Central Tendency   186
9.8   Variability   187
    9.8.1 Range   188
    9.8.2 Standard Deviation (Sigma)   188
9.9   The Normal Curve and Sigma   190
    9.9.1 Probability of an Event   195
9.10 Statistical View of the Process   197
    9.10.1 Process Control   197
    9.10.2 Process Capability   198

10 STATISTICAL PROCESS CONTROL CHARTS   208

10.1 Introduction to SPC Charts   209
    10.1.1 Concept of an SPC Chart   210
10.2 Control Charts and Variation   211
10.3 Developing a Variables SPC Chart   213

**11 INTERPRETING SPC CHARTS
FOR TEAM ACTION**   **230**

11.1 Why SPC?   231
11.2 Concept of Control   231
          11.2.1 Everyday Process Control   232
11.3 Process Control   233
11.4 Average versus Range   240
          11.4.1 Shifting Mean   240
          11.4.2 Increased Range   241
11.5 Out-of-Control Actions   242
11.6 Process Improvement   245
          11.6.1 Special Cause Corrective Action   247
          11.6.2 Common Cause Corrective Action   248
          11.6.3 Forcing a Process out of Control to Improve   250

**12 PROCESS CAPABILITY
AND IMPROVEMENT**   **258**

12.1 Process Capability   259
12.2 Process Control and Capability   261
12.3 Calculating Process Capability   262
          12.3.1 General Process Capability   262
          12.3.2 Calculating Process Capability—Centered
          Distribution   264
          12.3.3 Calculating Process Capability—Noncentered
          Distribution 267
          12.3.4 Process Capability and Improvement   270

APPENDICES   **277**

Appendix 1: Summary of Different SPC Charts   279
Appendix 2: Explanation of SPC Chart Patterns
                    and Possible Causes   287
Appendix 3: Standard Normal Probability Table   291
Appendix 4: Variables Statistical Process Control Chart   294
Appendix 5: Wafer Fab Terminology   296
Appendix 6: Flowchart of Team Actions for Low OEE   301
Appendix 7: Glossary   302

BIBLIOGRAPHY   313
INDEX   315

1

✳ INTRODUCTION TO MANUFACTURING

Manufacturing work is structured and repetitive, thus it appears easy to comprehend. If you consider a complete factory, however, the overall manufacturing line becomes complex—it's dynamic, full of people in motion, noise, and machines. Everything has a purpose. Whether you work in a factory or are visiting one, you may ask yourself, "What makes all this happen?"

If your studies are preparing you to work in manufacturing, then one day you will be a part of a manufacturing process. You want to avoid starting work lost in a big enterprise called production. As you proceed through this book, you will become more knowledgeable about manufacturing. You might also have a better idea of what your contribution could be to a manufacturing firm.

OBJECTIVES

After studying the material in this chapter, you should be able to:

1. Define manufacturing in terms of value, product, and process.
2. List how people work in manufacturing.
3. List and describe the four characteristics of high-volume manufacturing.
4. Describe the three criteria for a product to succeed in a competitive market, and explain how they are interrelated.
5. Calculate the manufacturing efficiency of an operation when given the actual and theoretical effort.
6. Understand the difference between value and waste in a process.
7. List and describe the four elements of manufacturing, and understand how they contribute to the resources needed to manufacture.
8. Describe manufacturing variables, their levels, and how they are set.
9. Understand how high-volume manufacturing applies to different products regardless of the technology in the product.
10. Describe why the market is important for manufacturing, and how the market can distort the manufacturing efficiency of a company.

1.1 WHAT IS MANUFACTURING?

Manufacturing is the process of adding **value** to a material to build a **product.** The act of manufacturing implies a **process,** or some repetitive sequence of **operations,** used to build the product. Manufacturing requires resources to produce the product, including an infrastructure of people in a firm providing the necessary support for manufacturing.

The concept of a manufacturing process is relative—it can be the total process (all operations and the necessary support activities) or a smaller subset of these operations, such as the process steps at a specific **workstation** centered on a piece of equipment. Different terms for a manufacturing process are shop floor or manufacturing line.

Value is added in manufacturing when work transforms a material into some functional form that is useful to humans. Adding value involves changing a material—shaping it, forming it, removing part of it, and so on. Manufacturing often has equipment that works on material to thus alter it in some manner. The material can be in the form of raw material (e.g., raw iron) or preprocessed material (e.g., purified silicon for semiconductor wafer fabrication). The work needed to add value is either physical or mental and often uses machines to extend the human capability.

People work in manufacturing in three ways: directly in the process, in support of the process (indirect), and in management of the process. These categories overlap, such as someone who works in the process also manages the activities at a

Table 1.1 How People Work in Manufacturing

In the Process	*Process Support*	*Process Management*
Operator	Shipping	Manager
Technician	Failure Analysis Lab	Management Staff
Maintenance	Accounting	Facilitator
Engineer	Parts Procurement	Coach
Quality Control	Training	Supervisor
Team Leader	Equipment Supplier	
Line Technician	(Vendor) Parts Supplier	

workstation, or a support person in the failure analysis lab could work in the process to identify how an equipment problem is related to a product defect. Examples of the different ways people work in manufacturing are shown in Table 1.1.

A manufacturing process is typically classified based on the number of products produced in the process, and is defined as either low volume or high volume (the former is also known as a job shop). Either type has unique characteristics about how value is added to the product. A low-volume process involves specialized work with a discontinuous flow of parts, such as building a special part for a government project. It is typically used for prototype work, early product development, or custom work. All studies in this book will be about **high-volume manufacturing.**

1.1.1 High-Volume Manufacturing

The characteristics of high-volume manufacturing are the same regardless of the product. The major characteristics are:

- *Product flow*
- *Standardization of work procedures*
- *Part interchangeability*
- *Repeatability of process*

Product Flow

Product flows through the different operations of a high-volume manufacturing process, similar to water flowing through different pipes in a water delivery system.

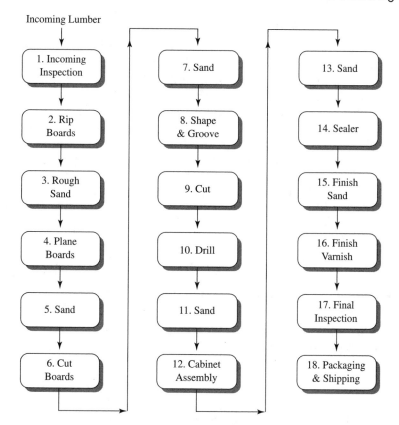

Incoming Lumber

1. Incoming Inspection	7. Sand	13. Sand
2. Rip Boards	8. Shape & Groove	14. Sealer
3. Rough Sand	9. Cut	15. Finish Sand
4. Plane Boards	10. Drill	16. Finish Varnish
5. Sand	11. Sand	17. Final Inspection
6. Cut Boards	12. Cabinet Assembly	18. Packaging & Shipping

Figure 1.1 Process Flow for Building a Wooden Cabinet

The product flow process may be described as continuous-flow manufacturing or just-in-time (JIT) manufacturing. Product flow exists in a high-volume process because the operations are interdependent. Each step in the process requires good incoming parts at the correct time, and supplies good parts to subsequent operations when needed. Process steps are linked and defined as either upstream or downstream from any particular operation in the process. A product flow for manufacturing wooden cabinets is shown in Figure 1.1.

Product flowing through a process is interpreted in different ways in manufacturing. Parts are ideally produced at a rhythmic interval with minimal parts stored at each operation. Parts are allocated a certain amount of effort at operations. A small error at a workstation can lead to several defects because of the product flow. People in the firm can appreciate how improving a high-volume process often occurs in the context of managing the overall product flow.

Standardization of Work Procedures

Work in a high-volume process is standardized at each operation. Standardization of work procedures means a particular work activity is repetitive and requires the same effort from all people across the different work shifts. This need for standardization in high-volume manufacturing derives from having similar product built to the same criteria, thus requiring repetitive steps to ensure reproducibility. Small procedural changes that are improperly implemented can lead to catastrophic results in a high-volume process.

   Each employee adheres to similar work procedures for the same operation. If a part is loaded into a tool in a certain manner, all operators follow the same steps to load the part. Manufacturing documentation is written to provide a comprehensive set of requirements for the job to be done correctly. Training ensures that all people follow the same standards, thus minimizing individualism and the desire to perform the work differently.

 EXAMPLE 1.1   STANDARDIZATION OF PROCEDURES

Give three common examples to illustrate how standardization of work procedures produces consistent work output.

Solution

1. An automotive repair shop gives free cost estimates to customers for repair jobs. Once the mechanic determines the source of the problem, it is then entered into a computer database. The computer calculates the number of hours and total cost to repair the automobile based on standard repair times. Thus, the cost estimate is the same no matter which mechanic does the repair.
2. Cookie cutters are used to make cookies with certain shapes from cookie dough. The cookie cutters standardize the cookie-making process to give the correct shapes.
3. A standardized number catalogues library books. Once a book's number is found in a centralized database, the book is then retrieved at a certain shelf location in the library. This system permits any person to find the same book.

Standardization in manufacturing requires discipline, which is often contrary to an individual's desire for uniqueness. Manufacturing discipline requires people to adhere to the existing work procedures, while maintaining individual initiative to improve these same work procedures. Thus, the need for structured, team-based manufacturing improvement supports the individual within a framework for group action.

BRAINSTORMING EXERCISE

Standardization

Refer to the process flow diagram in Figure 1.1. Brainstorm ways to standardize the work procedures to improve production of the wooden cabinet. These procedures may involve tools, fixtures, and documents to ensure reproducibility. Try to find procedures that apply to a high-volume process flow.

Part Interchangeability

Part interchangeability is critical in a high-volume manufacturing process. This characteristic means that in a group of parts, or lot, any part can be selected for manufacture because all criteria are within a specified **tolerance,** which is the basis for mass production. All functionally important product criteria are specified by tolerances. Parts become interchangeable when they are manufactured within tolerance as defined on a product drawing or specification. **Specifications** derive fundamentally from customer needs and are translated into product specifications by design engineers. The designer analyzes product requirements to ensure that tolerances are properly specified, so that the product will function correctly once built. Manufacturers use quality control techniques such as inspection, product test, and statistical process control (SPC) to ensure that products meet their specifications at different operations in the process.

Tolerances consist of a nominal dimension (the ideal value that the product designer specifies and the process strives to produce) and specification limits. The specification limits define an **upper-specification limit (USL)** and a **lower-specification limit (LSL).** Parts are called out-of-spec, or defective, if they are built outside of the specification limits. Note that building product to tolerances has other important aspects beyond interchangeability, such as ensuring correct product performance.

 EXAMPLE 1.2 INTERCHANGEABILITY

Refer to the wooden cabinet process flow diagram of Figure 1.1. How could part interchangeability be addressed for cutting boards in step 6? Explain how poor control of the board-cutting dimensions could lead to a problem later in the process.

Solution

The best way to ensure that boards are properly cut is to standardize the cutting procedure so that every board is cut with the same specified dimension from the product drawing. One method is to have a fixture on the table saw to accurately position

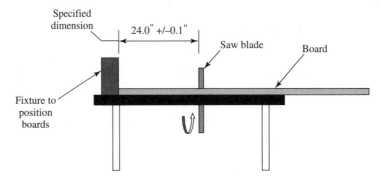

Figure 1.2 Table Saw with Board Stop

every board for the same cut, as shown in Figure 1.2. If this fixture is not used to accurately cut boards, then the dimension and tolerance for the board length could be out-of-spec, which creates problems during cabinet assembly in step 12.

Repeatability of Process

A high-volume manufacturing line requires that the process parameters of an operation be repeatable from process to process for similar products. Repeatability is important if multiple pieces of equipment and operators are running the same product. Examples of process parameters include actual machine settings or material properties such as density and purity.

If a process is not repeatable, then its output is not predictable. If the final product has poor performance for the customer, then it is of poor quality. Only a repeatable process can ensure high-quality product performance. Process repeatability is therefore a major goal for improving manufacturing.

 EXAMPLE 1.3   REPEATABILITY

Refer to the process flow diagram for making wooden cabinets in Figure 1.1. Consider the following steps in the flow:

Step 7: Sand
Step 9: Cut
Step 10: Drill

For each step, give three examples of process parameters that must be closely controlled to achieve repeatability.

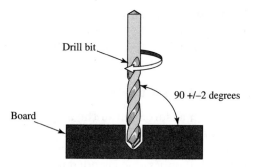

Drill bit

90 +/–2 degrees

Board

Figure 1.3 Drill Bit at 90 Degrees with a Tolerance of +/–2 Degrees

Solution

Step 7 (Sand):

    a. Grit of sandpaper (the roughness of the paper)
    b. Speed of sander
    c. Pressure on sander when used on the board

Step 9 (Cut):

    a. Sharpness of cutting blade
    b. Speed of cutting blade
    c. Number of teeth on cutting blade (fewer teeth give a rougher cut)

Step 10 (Drill):

    a. Sharpness of drill bit
    b. Rotational speed of drill bit
    c. Angle of drill bit (must be held at a nominal angle of 90 degrees within a specified tolerance, as shown in Figure 1.3)

1.2 COMPETITIVE MANUFACTURING

The final functional form of all value added in manufacturing is embodied in a product. Products compete for customers in markets. In a competitive market, products compete primarily on price, quality, and delivery. When price becomes the most important factor, then the market is termed a commodity market, which is the most advanced type of competitive market. Examples of products sold in competitive markets are disk drives and memory chips.

To succeed in a competitive market, a company must meet the following three goals in manufacturing.

- *Manufacturing efficiency (lowest cost)*
- *Manufacturing quality (highest quality)*
- *Manufacturing productivity (shortest delivery)*

Competitive manufacturing is the ability to build products that optimize these three goals. These goals will serve as the criteria for judging whether activity adds value for a manufacturing firm. Ultimately, they guide a firm while on the improvement path, positioning it to be competitive in the market.

Manufacturing Efficiency (Lowest Cost)

Manufacturing efficiency is a measurement of the amount of value added to build a product. *Efficient manufacturing lowers the cost to produce a product,* which meets the first criterion for a successful product in a competitive market. A process that is 100 percent efficient means that only value is added to produce the product, whereby only the necessary effort is expended in production. Note that such high efficiency does not typically occur in manufacturing, because most manufacturers expend unneeded effort to produce their product. Lowering manufacturing costs does not necessarily imply the product is sold inexpensively, as the selling price is set at the level the market can bear.

   Just because work is done in manufacturing does not mean that value is added to the product. Any time an effort is expended in manufacturing and that effort is not required for the product to function as intended, waste is added. Waste (also called non-value add) is the opposite of value, in that it is not needed and only increases the manufacturing cost of the product somewhere above its lowest possible cost. Substantial waste occurs in many manufacturing processes such as bad parts that must be scrapped, or idle equipment.

 EXAMPLE 1.4   VALUE VERSUS WASTE

Consider the following common processes, and explain what is value and waste with respect to high-volume manufacturing.

1. Attaching two parts manually with a screw and screwdriver.
2. Painting varnish on wood with a brush.

Solution

1. The act of turning the screw into the two parts is waste. The only value is the final turning of the screw that attaches the two parts, and the need for the screw to have a proper torque to hold the two parts over time. This waste in manufacturing could be replaced with an air-driven torque driver that turns the screw quickly, and a torque limiting device that properly seats the screw.
2. Applying varnish to protect the wood is value; however, using a spray painter instead of a brush will reduce the time needed to apply the varnish on the wood, therefore reducing waste. Uniformity of the varnish could be improved by using an automated spray booth, which sprays varnish in a repetitive sweeping action over a standard size object.

Efficiency in manufacturing is measured with respect to two efforts: the actual effort (what a firm expends today to produce) and the theoretical minimum effort (the minimum effort that could be expended when all available technology is considered). Manufacturing efficiency is estimated as

$$\text{Manufacturing Efficiency} = \frac{\text{Theoretical Minimum Effort}}{\text{Actual Effort}}$$

Time is a convenient measure of effort, such as the time it takes to perform an operation. Most manufacturing firms know their actual effort expended at a process step, because they are able to measure the time it takes to do the work. The difficulty in assessing manufacturing efficiency is determining the theoretical minimum effort required to accomplish the same task.

 EXAMPLE 1.5   ESTIMATING EFFICIENCY

To illustrate the concept of manufacturing efficiency at an operation, consider the sealer operation in the process flow diagram of Figure 1.1. A liquid chemical is applied to the finished cabinet to seal the wood. It requires a 45-minute cure at room temperature. A new sealant is developed which can cure in 10 minutes. What is the manufacturing efficiency?

Solution

Calculate the manufacturing efficiency as follows:

$$\text{Manufacturing Efficiency (\%)} = \frac{10 \text{ minutes}}{45 \text{ minutes}} \times 100 = 22\%$$

Given the available technology, this process is 22 percent efficient, which means only 22 percent of the work at this operation is value. The remaining 78 percent is waste. Another way of interpreting efficiency is to say this operation takes 45 minutes to do what could have been done in 10 minutes.

Manufacturing efficiency is calculated relative to the best possible theoretical minimum effort known in the industry, given the available technology. It is often difficult to determine this minimum effort because we tend to close our minds to new approaches. Other times we start looking and find a competitor's trade secret that is not public knowledge. Many companies simply accept their current effort as acceptable, choosing to not assess their manufacturing efficiency. This tactic is not good for manufacturing. A firm must seek as much legitimate information as possible about the market, competitors, equivalent product performance, and new manufacturing technology and use this information to estimate its own manufacturing efficiency (known as benchmarking).

In some circumstances the manufacturing process runs with poor efficiency to benefit another competitive goal, such as delivery. An example would be using an older, slower tool to produce excess product during peak demand periods. This overproduction permits the company to meet customer delivery schedules without investing in a new tool. As we shall learn, manufacturing is a continual trade-off between competing resources to find the optimum condition for cost, quality, and delivery—a condition that never remains constant because of work in the process. To be competitive in the market, a firm must strive for the optimum condition of cost, quality, and delivery, which rarely translates into the maximum condition for each individual goal.

Manufacturing Quality (Highest Quality)

Manufacturing quality occurs when the effort to build the product conforms to all requirements necessary for the product to function. Requirements are based on customer needs defined from the market. To attain the highest quality level, the goal is defect-free products.

Products that perform well in the marketplace are essential in a competitive market. A quality product is rarely achieved immediately when manufacturing begins. In most cases, process development involves a learning curve, which is using previously learned knowledge during an improvement activity. The knowledge gained from the learning curve is represented in areas such as skilled personnel, new equipment designs, and optimized variable settings in the process. While undergoing improvement, a manufacturing process must be able to identify and control defects until they are eliminated, usually through a form of quality control.

Delivering defect-free products in a competitive market requires commitment from all areas of the firm. Special recognition must be given to the manufacturing process area, because it is here that products are produced. When manufacturing is committed to quality, then all efforts are directed toward producing only high-quality product. In other words, the *source of manufacturing quality is the manufacturing process,* which permeates throughout the entire firm.

Efficient manufacturing is synonymous with quality manufacturing. If firms efficiently build their own product, then they also practice quality in their workplace. Unnecessary effort such as building defective products or reworking bad parts to make them good is not done in either an efficient or quality manufacturing operation.

 EXAMPLE 1.6   QUALITY CONTROL

Refer to the process flow diagram in Figure 1.1 for making wooden cabinets. Show where inspection points could be inserted in the process to control the quality of the components and final assembly for the wooden cabinets.

Solution

Inspection is most effective when done at the operation where the work occurs (to be discussed in more detail in Chapter 5). Each operator is responsible for the quality

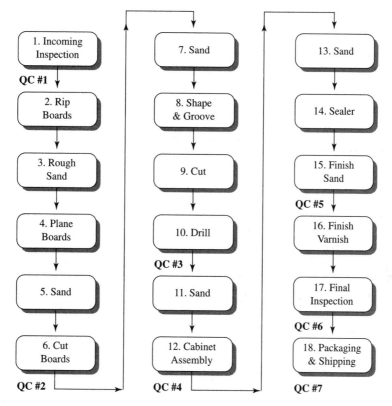

Figure 1.4 Process Flow for Building a Wooden Cabinet with Quality Control Inspections

of the product that is moved to the next downstream operation, which ensures that only good product moves through the process.

A process flow is shown in Figure 1.4 with the inspection points highlighted and numbered. The inspection criteria at each station are provided here. As a process improves, the number of inspection points and the frequency of inspection are reduced.

Quality Control Inspection Points

1. Inspect general incoming board quality (number of knots, flatness, etc.).
2. Inspect board thickness after planing, and length/width of cut boards.
3. Inspect shape dimension, depth/width of grooves, cut of board, and drill depth.
4. Inspect cabinet for overall dimensions, sturdiness, and cosmetic defects.
5. Inspect sand, sealer, and finish sand on cabinet.
6. Inspect varnish and final assembly.
7. Verify all product literature is present and product is packed properly.

Manufacturing Productivity (Shortest Delivery)

Manufacturing productivity must produce products with the shortest delivery time to ensure success. In today's competitive environment, a product that is not delivered within a market window can lose its competitive advantage. As technology rapidly changes, product life cycles shorten and the window of opportunity for introducing new products narrows. The product market window of some new products is now measured in months instead of years.

There is no doubt that manufacturing exists to produce. Anyone who has worked in manufacturing has felt the overriding pressure to meet production quotas or, rather, "make the numbers." This feeling leads to the manufacturing dilemma that *producing parts on time does not have to be synonymous with efficiency and quality.* It is possible for a company to expend effort to produce its product while short-changing quality and efficiency to get more parts shipped. This situation becomes real when a firm commits to delivering a product on time without making an equivalent commitment to defect-free, efficient production.

Largely emphasizing production quantity while ignoring efficiency and quality can damage a manufacturing firm. The product is shipped, but it may not work or it is too expensive. A firm must satisfy all three criteria for success in a competitive market.

1.3 ELEMENTS OF MANUFACTURING

The four basic **elements of manufacturing** that represent the manufacturing resources required to produce a product are:

- *People*
- *Methods*
- *Machines*
- *Materials*

These elements are the building blocks of a process, and together create an infrastructure that supports manufacturing both directly and indirectly. On one hand, understanding these different elements can help simplify the manufacturing concept. On the other, it is the interaction of these elements in all their various forms that makes manufacturing so complex.

Manufacturing elements are resources; therefore, the objective is to minimize dependency on these resources. The elements represent the specific aspects of manufacturing that should be used efficiently: fewer humans, less equipment, less time, and less material required. This efficiency increases the value in the product, which helps meet the three goals for competitive manufacturing.

People

People are the individuals who have specific roles in manufacturing. They include operators, technicians, shipping personnel, administrators, bookkeepers, and managers. Some people are directly involved in the manufacturing process, while others work in support positions such as facilities, purchasing, and accounting. All people in manufacturing interact through an organizational structure designed and created by management.

Methods

Methods include the procedures, instructions, and documentation that define how the manufacturing steps are carried out from suppliers through manufacturing to customer support. Procedures specify how manufacturing work is performed. They are described in documents such as process specifications, which specify the critical operating conditions used in the process. Documentation may be in software form (e.g., product recipes that are downloaded from a central software database) or in hardcopy form.

Methods also encompass the machine operating parameters, such as conveyor speed or furnace temperature, and process parameters, such as the yield of good parts. Another important aspect of methods includes consumables used to support production, including such items as gloves and tweezers. Much of the information for manufacturing methods originates as knowledge learned by technical people during the initial process development activity and is built on by other people during subsequent process improvement efforts.

 EXAMPLE 1.7   MANUFACTURING METHODS

Refer to step 2 of the process flow in Figure 1.1 for making wooden cabinets. Describe two different manufacturing methods that could be used during this step of the production process.

Solution

1. An example of a manufacturing method is the written procedure for setting up the table saw to rip boards in step 2 (ripping is the initial rough cut of the board). This procedure specifies the following process and equipment parameters:
   a. Safety procedures
   b. Type of boards that can be ripped (by board size and type of wood)
   c. Method for positioning stop for a particular type of board
   d. Height and rotational speed of the saw blade on the table saw
   e. Number of boards permissible to cut at one time
   f. Speed that boards are pushed through the table saw
2. The inspection document for boards ripped in step 2 is an example of a manufacturing method. This document has the following inspection criteria:
   a. Required nominal dimension and tolerance for all ripped boards
   b. Frequency of board inspections (e.g., every tenth board)

c. Method of measuring boards (e.g., with tape measure, go-no-go measurement fixture)

d. Unacceptable visual defects in ripped boards (e.g., major wood cracks along grain)

e. Necessary action when defect is found

Machines

Equipment is a major investment for modern manufacturing firms. An example of a technically advanced factory today is a semiconductor wafer fabrication factory (where semiconductor chips are fabricated). This factory can cost $1.5 billion or more, and the equipment investment is approximately two-thirds of the total capital cost. Given the large investment in equipment, the efficient use of equipment resources is important for competitive manufacturing.

Machines include the equipment, tools, and fixtures used in manufacturing to build and assemble the product, and the facilities necessary to support the equipment. Modern equipment is technically sophisticated, with a systems approach to integrating the machine functions through hardware and software interfaces. Advanced manufacturing equipment requires skilled operators and technicians to properly operate and support it for production.

 BRAINSTORMING EXERCISE

Equipment

Refer to the process flow in Figure 1.1 for manufacturing wooden cabinets. Brainstorm to find the different types of equipment used in the process. Separate the equipment into hand equipment and automated equipment. Discuss what type of operator and technician skills are required to support the different types of equipment.

Materials

Materials include all the raw and processed stock used to build the product. Process gases, chemicals, and incoming material are part of this element. Materials also include product material used during production, such as lumber for wooden cabinets.

 EXAMPLE 1.8   MANUFACTURING MATERIALS

Refer to the process flow diagram in Figure 1.1 for manufacturing wooden cabinets. Identify five different materials used in the manufacture of the cabinets.

Solution

Five materials used to manufacture wooden cabinets are (1) veneer plywood (plywood with a high-quality wood surface), (2) sealant, (3) varnish, (4) staples, and (5) glue.

1.3.1 Manufacturing Variables

Manufacturing variables are used to break down the four elements into different entities in the process. Variables are parameters that can be varied to specified settings based on the needs of the process to produce the part. Variables are typically specified in documented procedures so that the process settings are repeatable.

The term *variable* implies that the parameter value can be set at different settings, depending on what is found to be the best condition to manufacture the product. For instance, a machine voltage setting, or level, could be at 100 mV, 150 mV, or possibly 200 mV, depending on the particular process requirements defined in the documentation.

The setting of process variables is based on rigorous process evaluation to find the optimum setting for meeting the three criteria for competitive manufacturing. Many equipment and process parameter settings are determined by engineering in the early phase of process development, however; it is difficult to optimize these parameters because of the low volume of parts available. Optimization is usually achieved during product manufacture, and is best accomplished with the assistance of the people who work in the process—the operators and technicians.

To avoid guessing, the use of manufacturing methods helps to analyze a process and define the optimum variable settings to achieve process efficiency. Due to the difficulty of achieving process optimization, many manufacturing processes are not adequately evaluated to find the optimum variable settings. They are consequently operated at suboptimum conditions, which makes the firm less competitive in the market.

 EXAMPLE 1.9 VARIABLES

Refer to Example 1.3. The three process parameters found for the sand operation in step 7 of Figure 1.1 are:

   a. Grit of sandpaper (the roughness of the paper)

   b. Speed of sander

   c. Pressure (how hard the sander is pushed down on the board)

Evaluate different possible variable settings for each of these parameters at the sand operation.

Solution

   a. Grit of sandpaper: The variable is the grit number of the sandpaper, or the size and number of particles per unit area on the paper. For this process, the grit can vary between 20 grit (very rough) to 1000 grit (very smooth), in increments of 20 grit. This variable is specified in the procedure according to the type of sanding.

   b. Speed of sander: The variable is speed setting of the sander, which has three different settings: slow, medium, and high. The operator sets this variable based on which of the three settings is specified in the procedure.

   c. Pressure: The variable is the amount of pressure actually applied to the hand sander. This variable has no numerical or fixed setting. The procedure instructs the operator to use light sanding or heavy sanding, and each operator interprets this in the best manner possible.

1.4 MANUFACTURING AND THE PRODUCT

The high-volume manufacturing principles discussed in this chapter are the basis for understanding any product produced in volume. Although firms use the principles differently depending on their particular product, the basic manufacturing principles remain the same. A product can embody low technology, such as a hammer or plastic bucket, or represent high technology, such as a computer disk drive. In either case, if the product is produced in volume, then the common bond is the use of the manufacturing principles learned in this chapter.

Because the manufacturing concepts remain the same, we will now bridge from this simple product, a wooden cabinet, to a more complicated product, a semiconductor microchip. We will use the microchip fabrication process for examples and case studies throughout the book to highlight manufacturing. Both products are produced in volume. The semiconductor fabrication process employs more complicated product technology than a cabinet, and its market needs may require a different approach to manufacturing, but the fundamental manufacturing principles are the same as those learned for the wooden cabinet. As we move through this book, you can always return to this simple process flow to understand the principles.

1.4.1 High Technology—Semiconductors

Semiconductor manufacturing produces microchips on wafers of thin silicon. The microchips, or chips, are used in a wide range of modern electronic products. Silicon (a form of ultrapure sand) is used because of its special atomic structure, which makes it practical for constructing electronic devices (e.g., resistors, capacitors, and transistors) used on the chips.

Semiconductor manufacturing occurs in a factory known as a wafer fabrication facility. It is also called a wafer fab or fab. Many older wafer fabs are still in use; however, the equipment technology in a new fab remains state of the art for only about 2 years because the microchip product technology requires ever smaller sizes on the chip with more electronic functions to meet product specifications. This change in turn drives more complex equipment and additional process steps. Note that with respect to manufacturing, whether it is a tolerance of +/– 0.125 inches for a wooden cabinet or +/– 50 angstroms for a microchip, the need to control dimensions and tolerances to meet a specification does not change. This principle is fundamental. It is how we apply our effort to actually control the tolerances that will be different, not the need for the tolerances.

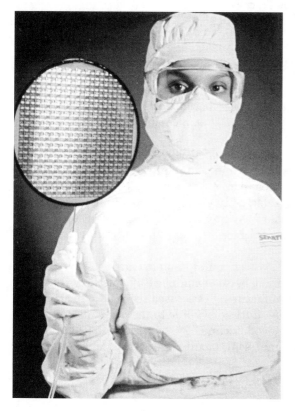

Operator in cleanroom attire
with a semiconductor wafer.
Photo courtesy of Sematech Archives.

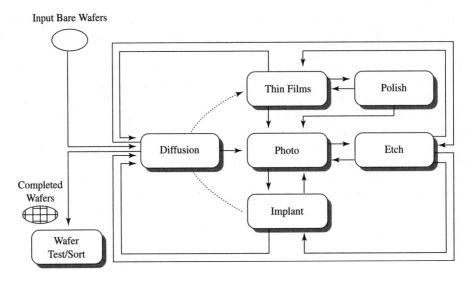

Figure 1.5 General CMOS Process Flow in a Wafer Fab[1]

1.4.2 Semiconductor Process Technology

A short overview of semiconductor process technology is in order. This overview will be sufficient to discuss manufacturing, not to become an expert in semiconductor manufacturing. You can learn more about the semiconductor process in books that address this subject.

In semiconductor chip fabrication, the process consists of building up many layers of thin materials deposited on the silicon wafer and requires silicon wafers to cycle through certain operations many times (e.g., a wafer may pass through one operation 15 to 20 times). The layers are selectively modified through an etch process (which could be compared with the board cutting and planing operations in the wooden cabinet process flow). For the major process areas in a wafer fab, refer to Figure 1.5 for the general process flow model.

This visual process flow model for a silicon wafer demonstrates the interactive movement of wafers in a wafer fab. Do not become confused by this flow diagram. Wafer fabrication is similar to all manufacturing processes (such as fabricating a wooden cabinet), during which the use of effort adds value to the basic material. The wafer fab process employs more advanced technology, but the need behind the process areas are the same (e.g., add a material, modify its surface, and join it with another material).

---

[1]See Reference 13 in the bibliography.

Table 1.2 Description of Major Semiconductor Process Areas

Diffusion:	The furnace process used for uniformly heating the wafer for any process requiring high temperature (up to 1200°C). The purpose could be to grow an oxide layer that acts as a dielectric layer for a capacitor. Another furnace process is to anneal, or restore, the wafer surface following other processes.
Photo:	Also termed photolithography and patterning, it is the process whereby circuit patterns from a photomask are transferred to the top surface of the wafer by way of an image transfer onto a light-sensitive resist material on the wafer. The resist material only permits certain areas of the wafer to be etched, which gives rise to the circuit pattern on the wafer.
Implant:	The ion implantation, or introduction, of selected impurities (dopants) into the silicon wafer surface by means of high-voltage ion bombardment. These ionic dopants (meaning the atoms have a positive or negative charge) interact at the atomic level with silicon atoms. Ion implant is how silicon achieves its particular semiconductor properties.
Thin films:	The location where very thin layers (films) of material are deposited on the wafer surface. It is called metallization if the deposited thin film is a metal (such as aluminum or titanium). The metal layers will eventually become the conductors within the semiconductor (just like electrical wiring in a house). The thin film process area also includes the deposition on nonmetal films such as insulators or plasticlike materials such as polyimide.
Etch:	A process for removing material in selected areas on the wafer surface. Etching a metal film will leave the unique circuit wiring to attach the different devices on the wafer surface. Etch could be compared with cutting or shaping in the cabinet process.
Polish:	A process for flattening (planarizing) the wafer surface, and is needed to smooth multiple layers on a wafer, as with sanding on a wooden cabinet.
Wafer sort:	The major test operation after wafer fabrication during which the electrical function of each chip on a wafer is tested. This operation "sorts out" the good chips from the bad chips on a wafer.

As seen in Figure 1.5, wafers cycle multiple times through major process areas. The terms are defined in Appendix 5. It is not important to understand all the technical details of this process to study manufacturing, but a short definition of each major area is given in Table 1.2.

Operator loads a cassette of semiconductor wafers in an etch tool.
Photo courtesy of Sematech Archives.

The process steps in Figure 1.5 have the same basic needs for high-volume manufacturing that were analyzed for the wooden cabinet in this chapter. The characteristics of high-volume manufacturing for a wafer fab are shown in Table 1.3.

Microchips produced from a wafer fab compete in a competitive market just as
wooden cabinets compete in their respective market. The strategies to compete in the
markets are different, but the competitive goals to succeed in the market are the
same. The firm that produces its product with the lowest cost, highest quality, and
shortest delivery has the highest probability of success.

1.5 MANUFACTURING AND THE MARKET

High-volume manufacturing produces products that compete in a competitive market. The market is the key factor that defines a manufacturing firm's need to be competitive. If the customer requires low-cost, high-quality products delivered at the
right time, then the manufacturing firm that best meets the requirements will stay

Table 1.3 High-Volume Characteristics of a Wafer Fab Process

Product flow:	Wafers flow through the manufacturing process, cycling through the different process areas many times. This pattern is similar to lumber flowing through the cabinet-making process, passing multiple times through process steps such as cutting.
Standardization:	Standardized work procedures across all equipment and employee work shifts are critical to successfully manufacture silicon wafers within the specified criteria. This standardization is the same as demonstrated for the cabinet process.
Interchangeability:	Silicon wafers and other materials in the wafer fab process are built to specifications with controls to ensure quality control, just as with the cabinet-making process. Interchangeable product is needed to successfully build multiple wafers with the same wafer fab process.
Repeatability:	Repeatability from tool to tool is necessary for the wafer fabrication process. This requirement is identical to the needs of the cabinet process.

in business. In this sense, it is the market that drives a firm's need to improve manufacturing and stay competitive.

Market conditions can actually permit a firm to be an inefficient manufacturer, yet still be profitable. An example of this is a firm that is the sole producer of a product. Another example is a firm whose product is protected from competition through government regulation. In these instances, customers will probably accept any product condition at the time of sale. The firm will then use this false market dominance to portray itself as efficient.

Financial market indicators such as profit, stock value, or return on investment (ROI) do not indicate whether a firm is an efficient manufacturer. They indicate how a firm is performing in the market. To assess whether a firm is efficient in manufacturing, analyze the process to determine the amount of value added to the product during manufacture.

Manufacturing is ultimately a business. Any person working in manufacturing contributes to the success or failure of the business, and thus is forced to be knowledgeable of the business environment in which that firm competes. This fact is true for all products, whether low or high technology. On the other hand, the converse of this is not true. A person who works in a business is not necessarily working in manufacturing (e.g., someone in sales). A businessperson may lack the knowledge to succeed in manufacturing.

SUMMARY

Manufacturing is the process of adding value to a material to build a product. People in manufacturing either work in the process, support the process, or manage the process. A high-volume manufacturing line has four major characteristics: its product flow, standardization of work procedures, part interchangeability, and repeatability of process. To succeed in a competitive market, a manufacturing firm must meet three goals: manufacturing efficiency (lowest cost), manufacturing quality (highest quality), and manufacturing productivity (shortest delivery). The four elements of manufacturing are people, methods, machines, and materials and are the basis for the variables found in manufacturing. The principles of high-volume manufacturing apply to any product produced in quantity, regardless of the technology in the product. The market is the key factor that defines the need for a manufacturing firm to be competitive.

IMPORTANT TERMS

Manufacturing	Specifications
Value	Upper Specification Limit (USL)
Product	Lower Specification Limit (LSL)
Process	Repeatability
Operations	Competitive manufacturing
Workstation	Manufacturing efficiency (lowest cost)
High-volume manufacturing	Manufacturing quality (highest quality)
Product flow	Manufacturing productivity (shortest
Standardization of work procedures	delivery)
Part interchangeability	Elements of manufacturing
Tolerance	Manufacturing variables

REVIEW QUESTIONS

1. What is manufacturing?
2. How is value added in manufacturing?
3. What are the three ways people work in manufacturing?
4. List and describe the four major characteristics of high-volume manufacturing.
5. What are the three criteria for a manufacturing firm to succeed in a competitive market, and how are they interrelated?
6. What is waste and how is it related to value?

7. Define manufacturing efficiency.

8. What are the four elements of manufacturing? Give an example of each.

9. What is a manufacturing variable, and how is it important for optimization?

10. Describe the relationship between manufacturing and the market.

EXERCISES

Repeatability of Process

1. Refer to the process flow diagram in Figure 1.1 for making wooden cabinets. Based on the information provided, brainstorm to find possible process parameters that are important for repeatability of the process to produce high-quality cabinets. For three of the given process steps, make a list of three different parameters for each step. Example 1.3 already gives parameters for steps 7, 9, and 10.

Manufacturing Efficiency

2. Consider the process flow diagram in Figure 1.1 for making wooden cabinets. Brainstorm production techniques that could increase the manufacturing efficiency at the different process steps. The new techniques should lower the cost of the cabinets. An example is replacing a nail and hammer operation in assembly with a stapler and glue to reduce the assembly time. Another example is stacking boards during the cut operation (cutting more than one board), but be careful to not let an improvement in efficiency lead to problems in another area, such as poorer quality.

3. Explain how waste occurs in the following manufacturing situations, and how the process can be improved to reduce waste.

   A. A manufacturer of wooden cabinets has burrs on the edge of the board after the cutting operation. Burrs occur due to a dull cutting blade. Instead of sharpening the blade when required (this takes about 15 minutes per shift), the operator sands the edges. Sanding takes about 10 seconds per board, and removes most of the burrs.

   B. An operator performs a quality measurement on a group of semiconductor wafers when they arrive for processing. The wafer measurements are then entered into the computer database by typing the values at three different menu locations on the computer. It takes about 5 to 10 seconds per menu to input the data.

   C. A material is baked for 1 hour prior to processing. The bake step was added about 6 months ago when the product first started in manufacturing, because parts were being stored for up to 48 hours prior to processing. Baking reduced the moisture content of the parts. Process flow improvements have reduced the part storage time to a maximum of 30 minutes.

4. A production tool requires 4 hours to install a process kit after every 2 weeks of production. A new tool on the market requires 30 minutes to install the same process kit.

   A. Calculate the manufacturing efficiency of the existing tool.

   B. The firm has excess demand for its product. The firm purchases the new tool and installs it. What do you recommend the firm to do with the tool that has lower manufacturing efficiency?

Manufacturing Quality

5. Two operators work together at a furnace workstation, where wafers are exposed to high furnace temperatures to grow surface oxides. Both operators have 1 year of experience, and have undergone training and certification to work with the furnace equipment. Groups of wafers processed at the furnace workbay have specific processing procedures, and all operators have been trained to follow the same procedures.

   While working, one operator notices that the other operator is not following the correct procedure for maintaining cleanliness of the wafer boats (a type of fixture) used to hold the wafers in the furnaces. By reducing the amount of cleaning done to the boats, the operator is able to increase the number of wafers produced in a shift.

   A. Do you think it is acceptable for a person to modify the procedures to increase the number of parts produced? Explain your answer.

   B. Explain the potential consequences to product quality of modifying procedures with no analysis or approval.

Manufacturing Elements and Variables

6. Refer to the process flow in Figure 1.4 for building wooden cabinets. Consider each process step, and assign people to work at each step (or combination of steps). If you assign more than one step per person, explain how this will work. Include the need to have quality control. The goal is to have the minimum number of people needed to build the cabinet. Present your results with a process flowchart that shows where the people work and their responsibilities.

Manufacturing and the Market

7. Give an example of a manufacturing firm that competes in the following conditions:

   A. Fiercely loyal customer base that will buy its product under any conditions.
   B. The firm provides a product at a set price with minimal competition.
   C. Extended product base that makes it difficult for customers to buy elsewhere.
   D. Foreign competitors cannot easily compete against the product due to market conditions.
   E. A profitable manufacturing firm that does not manufacture its own products because it would not be competitive.

2

A SHORT HISTORY OF U.S. MANUFACTURING

This chapter gives background information for understanding manufacturing. We gain valuable insight into the current issues of manufacturing by studying its recent history. Understanding the past provides knowledge to analyze current manufacturing practices, with the aim of avoiding mistakes and repeating successes. This knowledge is the basis for future improvements.

OBJECTIVES

After studying the material in this chapter, you should be able to:

1. List people involved in early U.S. manufacturing and discuss their contributions to manufacturing process development.
2. Describe how manufacturing firms addressed capacity and quality in the 1950s and how this was related to the market; define capacity production.
3. Define the following terms: economic order quantities, economies of scale, and acceptance quality levels; describe how these models supported capacity production.
4. Describe the trends in technology, marketing, and U.S. manufacturing that occurred in the 1960s; explain how they affected American manufacturing expertise.
5. Explain what happened to the ability of U.S. manufacturing to compete in the 1970s, and give the primary reason why this occurred.
6. Describe the improvements in manufacturing that took place in the 1980s, and the new manufacturing techniques that were introduced. Identify the source of these new techniques.
7. Describe the positive and negative aspects of present-day trends in manufacturing improvement.

2.1 PRE–WORLD WAR II

Modern manufacturing originated from the industrial revolution of the 1700s and 1800s. Credit for the first U.S. system of manufacturing with interchangeable parts is given to Eli Whitney in 1780, when he received a contract for manufacturing 10,000 muskets.[1] Modern principles of manufacturing were developed in the later 1800s and early 1900s in such enterprises as the cotton mill, with a concentration of workers and raw materials to create the need for a factory system.

Frederick W. Taylor studied time-motion production methods in the late 1800s and assessed how to find the most efficient ways to perform work. He then **standardized procedures** for all workers, which reduced costs and improved quality. In the automotive industry, the assembly line was established in the early 1900s at Ford Motor Company, with **continuous flow** of the product during assembly. Henry Ford emphasized the identification and elimination of waste to be more efficient— an improvement later studied by Japanese companies.

Walter Shewart at Bell Laboratories introduced process control methodology into manufacturing with the development of **statistical process control (SPC)** charts in the early 1920s. This methodology permits prediction of process problems, and is the same as that used today by many companies. Statistical process control is discussed later in Chapters 9–12.

---

[1] See Reference 2 in the bibliography.

2.2 1950s: CAPACITY PRODUCTION

In the post–World War II international economy, U.S. industry had important access to commercial markets in Western Europe and Asia. To exploit these large markets, U.S. manufacturing firms focused on **capacity production,** which meant producing as many products as possible to meet demand.

The quality of U.S. goods was acceptable. Design methods were based on wartime criteria, when parts were intentionally overdesigned to ensure proper function. Companies had little interest, however, in manufacturing improvements that would increase product quality. W. Edwards Deming, an American, attempted to teach quality techniques to U.S. firms using statistical process control, but was rebuffed. Deming was invited to Japan to lecture on quality control in 1950, followed by Joseph M. Juran in 1954, and they continued to educate Japanese firms on quality. This quality training was said to have contributed to Japanese post–World War II manufacturing improvements.[2]

Universities supported capacity production by developing manufacturing models based on maximum capacity. The **economic order quantity (EOQ)** model for manufacturing emphasized large lot sizes for production and ways to hold inventory most efficiently. Process efficiency was analyzed with **economies of scale,** which strives to maximize the quantity of parts moving through a process. The economies of scale model has validity when used wisely (for example, getting more people to ride in a single car is more efficient than each person driving alone), but was overemphasized at the time to the detriment of other important needs for manufacturing improvement.

2.3 1960s: TECHNOLOGY AND MARKETING

In the 1960s, U.S.-produced goods faced increased competition from international companies, primarily from Japan. The Japanese were entering new markets with low-technology products to develop their manufacturing base, with manufacturing products for U.S. companies who later put their own name on the end product. During that time, foreign goods were often belittled, and the United States was believed to produce superior high-technology products.

Companies in the United States neglected growing manufacturing problems at home; the result was a decrease in product quality and manufacturing expertise. Quality in the manufacturing organization centered on final inspections to find bad parts, with the use of *statistical models to justify accepting incoming defects.* These models were based on a technique known as **acceptance quality levels (AQLs).** This tech-

---

[2] See Reference 10 in the bibliography.

nique is still used in some firms today. In the AQL model, product is accepted from a supplier as long as defects are kept under a certain statistical level. For example, a 5 percent AQL will accept 5 percent defects in a group of parts (on the average). Universities dropped manufacturing as a discipline in their engineering and technical schools. Instead, they studied methods to optimize manufacturing for larger capacity. An example is **material requirements planning (MRP),** an inventory control method, based on large production lots, that complements economic order quantity and economies of scale.

Industry's strategy was to emphasize **research and development (R&D)** and marketing, but in the process ignored manufacturing. This excessive focus on technology led to the deterioration in quality of manufactured products.

2.4 1970s: RETREAT

International firms, primarily from Asia, were serious in their drive to enter and control U.S. markets. U.S. industries lost market share in products from raw steel production and automotive, consumer electronics, machine tools, and semiconductor manufacturing. Competitors were successful because their products were efficiently manufactured at lower costs and higher quality. Industries in the United States, however, did not acknowledge foreign manufacturing expertise, explaining their successes as only an improved copy of what this country did (known as **reverse engineering**).

Manufacturing in the United States was in turmoil, having lost critical manufacturing knowledge during the capacity production era. Unable to correct the problem, management reacted by closing many U.S. manufacturing plants and moving production to countries with low-cost labor. The benefits from this retreat on manufacturing were immediate, with lower manufacturing costs, increased quality due to developing overseas manufacturing knowledge, and management removed from the nightmare of trying to improve U.S. manufacturing problems. This change in strategy led to the concept of **"black box" production:** raw materials went into a "black box," manufacturing value was added to the materials, and a finished good was delivered. The black boxes, in this case, were developing countries that imported raw materials and developed their manufacturing base to produce goods for U.S. consumption.

Sending manufacturing work overseas complemented the emphasis in the United States toward research and development (R&D) and marketing. The U.S. strategy was for companies to design and develop the product, send it abroad for manufacture, and then market it in the United States. The justification was that manufacturing was low technology and thus not needed. This mindset accelerated the shift in the United States from a manufacturing to a service economy (restaurant, insurance, etc.). Not considered at the time, however, was the subsequent lower pay, menial work, and limited individual growth potential for many people in the service industry.

2.5   1980s: STANDING UP

By the 1980s, no one could deny that U.S. productivity had serious problems, with plants closing, the subsequent loss of jobs, and the inferior quality of many U.S.-made products. It was difficult, however, to face the problem's root cause of lost manufacturing expertise after so many years of neglect.

The initial reaction of many industrial leaders was to buy their way out. Based on U.S. high-technology emphasis, many firms started major programs to automate factories and hopefully improve productivity. The United States, however, quickly became ensnared in the flaw of constructing an automated, "high-tech" manufacturing factory, and ignored the lack of knowledge about basic manufacturing methods.

Enlightenment to improve U.S. manufacturing processes came from an unlikely source: Japan. **Japanese manufacturers** published books and gave lectures to U.S. manufacturers on how they had successfully developed new manufacturing techniques after the war. They gave detailed information on how to improve manufacturing based on such concepts as continual improvement, zero defects, and quality teams.

With this new knowledge source for **manufacturing improvement,** an effort began across diverse U.S. industries to understand the manufacturing problems. Industries learned to view manufacturing as a system with interaction between the elements and to integrate manufacturing into business activities. Many firms embraced manufacturing teams, emphasizing self-sufficiency coupled with communication between the broad disciplines required to support manufacturing. Over time, the quality of U.S. products started to improve.

American industries learned to work together to find solutions to their problems. An example is the formation of Sematech in the mid-1980s, the semiconductor consortium for manufacturing research and development. Half of Sematech was funded by member companies and the other half by the Department of Defense, which considered the U.S. manufacturing problem so severe that it became an issue of national defense.

2.6   1990s: THE CROSSROADS

Modern global corporations have no national boundaries. There is less of an illusion of "others" with their incompetence, and "us" with our superiority. Manufacturing is the basis for many developing nations to compete in the world market. At the same time, advancing countries now have to deal with social, monetary, and other types of problems associated with rapid economic advancement.

Manufacturing improvements continue in many U.S. firms, with substantial knowledge gained from the learning curve of earlier problems (such as the haphazard equipment automation in the 1980s). Management gives more support to manufacturing as an integral part of the firm's business strategy. Employees recognize the need to be flexible and cross-trained for multiple tasks.

At the same time, **manufacturing excesses** proliferate. Highly paid consultants offer any number of solutions to a problem, but do not stay to see if their ideas work. Companies portray their manufacturing processes as efficient and advanced, using marketing techniques and often stretching the truth to justify their claims. Consultants dictate to firms what is required to win awards, while management teams spend an inordinate amount of time and money preparing for presentations and audits. The negative consequences of this marketing of manufacturing are skepticism for valid improvement ideas and eventual confusion and low morale among employees. Why? The people working in manufacturing are not involved in defining the problems and solving them.

Industries have been largely profitable in this decade, which leads to the illusion of efficiency. People downplay such benefits to manufacturing as the weakened dollar, which handicapped international competitors for competing on a cost basis in the early 1990s. Many manufacturing jobs that exist today are merely final assembly and test lines, with most of the manufacturing value added during the production of components in offshore factories. American firms have laid off large numbers of people, who are then reabsorbed into the workforce in nonmanufacturing jobs in the service industry, usually at a lower salary and with less job security. The future of manufacturing in the growing service economy is at the crossroads.

SUMMARY

Modern manufacturing originated during the industrial revolution of the 1700s and 1800s, with significant advancements in the early 1900s for understanding how to efficiently manufacture products. Following World War II, U.S. manufacturers focused on producing the maximum number of products possible (capacity production), while ignoring quality and efficiency. As firms lost basic manufacturing knowledge, they replaced it with an emphasis on technology and marketing. By the 1970s, this strategy caused U.S. manufacturers to lose market share to competitors and eventually forced them to close factories and move overseas in search of lower wage costs. In the 1980s, U.S. firms began addressing their manufacturing problems, which initiated the start of basic improvements in their manufacturing practices. This improvement activity has also created manufacturing excesses that are detrimental to improvement.

IMPORTANT TERMS

Standardized procedures	Economic order quantity (EOQ)
Continuous flow	Economies of scale
Statistical process control (SPC)	Acceptance quality level (AQL)
Capacity production	Material requirements planning (MRP)

Research and development (R&D)   Japanese manufacturers
Reverse engineering   Manufacturing improvements
Black box production   Manufacturing excesses

REVIEW QUESTIONS

1. Describe how the following people contributed to manufacturing development: Eli Whitney, Frederick W. Taylor, Henry Ford, and Walter Shewart.
2. What is capacity production, and when did it develop in the United States? What effect did it have on manufacturing quality?
3. What are economic order quantity (EOQ) and economies of scale?
4. What happened to U.S. manufacturing quality during the 1960s? What are AQLs and how did they affect U.S. manufacturing quality?
5. What was the U.S. response to foreign competition in manufacturing during the 1970s?
6. How did U.S. industry finally gain knowledge to start improving manufacturing in the 1980s?
7. Give a short description of the general status of U.S. manufacturing in the 1990s.

3

MANUFACTURING IMPROVEMENT PROGRAMS

When U.S. firms revived their interest in manufacturing during the 1980s, they initiated many manufacturing improvement programs. This chapter is a survey of the major programs, their origins, how they relate, and their pros and cons for manufacturing. This chapter will also help those who are entering manufacturing to assess the potential contribution of a program to their workplace.

OBJECTIVES

After studying the material in this chapter, you should be able to:

1. Define zero defects, quality circles, and continual improvement, and describe how these programs are used to improve manufacturing.
2. Describe TQC and how this is applied to manufacturing improvement.
3. Define SPC and how it is used in manufacturing.
4. Define the terms JIT and kanban, and show how these are used in manufacturing.
5. Discuss how the 5S program and CANDOS are used in manufacturing.
6. Describe TPM, and its approach to equipment maintenance in manufacturing.
7. Explain TCS and how it defines customers in an organization.
8. Discuss the 6-sigma program and how it is used for quality improvement.
9. Explain TQM, its primary focus, and how it is implemented.
10. Describe ISO 9000, and discuss the primary manufacturing areas affected by this program.
11. Explain the two factors that determine the effectiveness of improvement programs.

3.1 IMPROVEMENT FROM JAPAN

Learning improvement from Japanese manufacturers, and the necessary Japanese terms and culture, has not been an easy task for American industries. After several years of grappling with notions of cultural superiority and ego, they eventually found common ground by realizing that all cultures have strong and weak points with respect to manufacturing improvement.

The improvement programs developed or used by Japan and brought to the United States are:

- *Zero defects*
- *Quality circles*
- *Continual improvement*
- *Total quality control (TQC)*
- *Statistical process control (SPC)*
- *Just-in-time (JIT) manufacturing*
- *5S manufacturing*
- *Total preventive maintenance (TPM)*

Zero Defects

The zero defects quality improvement program establishes a framework for not accepting any defects in manufacturing. There is *no acceptable level of defects* in a firm

37

that manufactures to zero defects, and any defect that occurs must be identified, analyzed, and have corrective action in place to keep it from occurring again.

The zero defect approach to manufacturing contrasts sharply with the American quality model, acceptance quality level (AQL), discussed in Chapter 2. The zero defect model requires corrective action to eliminate defects, whereas the AQL model accepts a certain percentage of defects in the process (as previously described, a 1 percent AQL means that a shipment of parts is accepted if, on the average, 1 percent or less of the products are defective).

Zero defects was initially belittled by many U.S. firms who did not understand its basic premise for continual improvement, and believed it was impractical and statistically impossible to expect zero defects. It is now recognized as a comprehensive approach that is as much a way of life as it is a set of practical procedures to achieve defect-free product.

Quality Circles

The concept of quality circles was first brought to U.S. manufacturing in the late 1970s. Other names have been used, such as quality improvement teams (QITs). Quality circles involve all production people in making decisions that affect manufacturing improvements. This decision making is done through structured team meetings, during which each person is encouraged to provide ideas for improvement. Quality circles were the first step toward manufacturing teams in the United States. They stressed the importance of seeking input for decisions from all production personnel.

Continual Improvement

Continual improvement is based on the belief that we should never stop trying to improve the process. It promotes improvement actions in small, incremental steps over time, building on existing knowledge to minimize risk from change. It requires a continual effort of all employees to question the status quo and find a better method to accomplish a task.

Total Quality Control (TQC)

Total quality control (TQC) is a term initially used by A. V. Feigenbaum as the title of his book published in the United States in the 1950s (*Total Quality Control* by Armand V. Feigenbaum, 1951). The concept of TQC was further developed by Japanese manufacturers to address all areas of manufacturing for making parts right the first time, with the responsibility for quality resting at the operation where the part is made. TQC is based on knowledgeable teams responsible for making production decisions. An overview of how TQC addresses manufacturing improvement is shown in Table 3.1.[1]

A review of the TQC concepts in Table 3.1 illustrates basic tools for controlling quality at the workplace. A fundamental part of TQC is its practice at all organizational levels of manufacturing, providing a system of practical techniques and idea sharing between everyone involved in the process. This concept is the basis for *total* quality control.

---

[1] See Reference 10 in the bibliography.

Table 3.1 Total Quality Control: Categories and Concepts

TQC Category	TQC Concept
Organization	• Everyone's responsibility is production.
Goals	• Make it a habit to improve. • Achieve perfection.
Basic principles	• Control the process. • Make quality easy to see. • Insist on compliance. • Stop production for problems. • Correct one's own errors. • Check every part for quality.
Facilitating concepts	• Use project-by-project improvement. • Use quality control to facilitate problems. • Move parts through process in small groups. • Keep the workplace clean. • Keep some equipment capacity as backup. • Do all daily checks on your machine.
Techniques and aids	• Highlight problems. • Find ways to keep problems from reoccurring. • Carefully inspect the first and last parts. • Use process analysis tools. • Use the team approach (quality circles).

Statistical Process Control (SPC)

Statistical process control (SPC) is a process analysis tool for collecting data about a process. These data are used to statistically control the process and make predictions about future process performance. Types of data used in SPC can vary, including equipment performance data, product measurements, and results from product testing. SPC permits easy data analysis, because it employs a visual format that shows the performance of the measured data over time. SPC will be covered in depth in Chapters 9–12.

SPC was originally developed in the United States in the 1920s as a tool to control and predict variation in manufacturing. It was successfully used by Japanese industries during their post–World War II improvement activities in manufacturing, and was brought back to the United States in the 1980s (with the development of process capability).

Unfortunately, SPC is used improperly in many U.S. firms. The problem is that SPC is fundamentally a statistical technique, and can easily be modified so that any process can appear good even though no actual improvement is taking place. When this happens, it is transformed into a marketing tool ("look at how well we control our processes with SPC"), used to imply that the firm has control over its process.

Just-In-Time Manufacturing (JIT)

The just-in-time (JIT) manufacturing system optimizes product flow in a high-volume production line by having the necessary parts arrive at a process only when they are needed. In other words, product is built at a particular operation based on demand from the downstream operation. This method results in reduction of the number of parts in a process as production cannot overbuild. Because JIT keeps process steps from overbuilding, the flow of parts throughout the process is balanced. Each operation must consider the number of parts needed at the next operation prior to building.

In a JIT production system, part flow in a process is controlled by use of **kanbans,**[2] a Japanese word meaning "visible record," which represents a product flow communication system between operations in a manufacturing line. With kanbans, operations are linked with some form of flow-control methodology (examples are racks that only accept so many parts, or lights that indicate it is alright to build). Kanbans serve to effectively highlight operations that restrict product flow in a manufacturing line, because parts back up in front of the operation (just like water backing up in front of a dam). This puts focus on problem areas where improvement is needed.

JIT also incorporates many organizational aspects of improvement, including teams and interaction between people from different areas of manufacturing. American companies have implemented the principles of JIT under different names, such as continuous-flow manufacturing, with varying levels of success.

5S Manufacturing

5S stands for five different Japanese words (all beginning with the letter s), which address operator discipline and control in the manufacturing workplace.[3] The words translated into English are *clearing up, arrangement, neatness, discipline,* and *ongoing improvement.* The first letters of these terms form the basis for the acronym **CANDOS** (adding the letter s for *safety*), which has been supported by Sematech in its total productive manufacturing program.[4]

The 5S program recognizes that no amount of organizational support for improvement can replace the fundamental need for discipline at the workplace. If the workplace is out of control, then the product produced at the workplace will also be out of control. The best way to organize a common workplace is with orderly procedures, which provide a standardization philosophy that can be used throughout manufacturing.

---

[2]See References 9 and 12 in the bibliography.
[3]See Reference 8 in the bibliography.
[4]See Reference 1 in the bibliography.

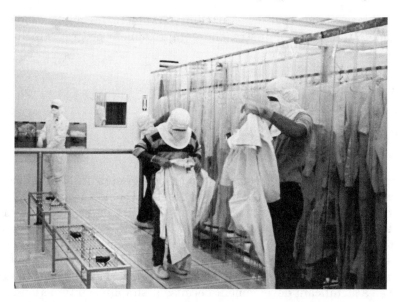

Manufacturing discipline requires adherence to all procedures, including the preparation to enter the work area.
Photo courtesy of Sematech Archives.

Total Preventive Maintenance (TPM)

Total preventive maintenance (TPM) is a system used to achieve optimum equipment performance. It was developed and implemented in the early 1970s in Japan.[5] TPM advocates a comprehensive maintenance program that strives for zero equipment failure, which serves to reduce product defects and delay the time needed for older equipment to wear out. A nonmanufacturing example of a piece of equipment used well beyond its expected life is the B-52 airplane. Its success was achieved by adherence to strict maintenance procedures.

The TPM program goes beyond the most basic maintenance activity of having technicians react to problems after they occur, because a tool that has gone down has already adversely affected manufacturing. It advocates team-based involvement in maintenance, where all employees in a firm contribute. Operators are viewed as a firm's first line of maintenance, supporting equipment through daily cleaning, regular preventive maintenance (PM) procedures, and team problem solving.

3.2 IMPROVEMENT FROM U.S. FIRMS

American firms developed various improvement programs during the 1980s, some of which were based on Japanese techniques, whereas others originated from U.S.

---

[5]See Reference 8 in the bibliography.

culture and ideas. Some programs started in manufacturing applications and were later applied to the service industry. These improvement programs include:

- *Total customer satisfaction (TCS)*
- *6-sigma program*
- *Market-driven quality (MDQ)*
- *Total productive manufacturing (TPM)*

Total Customer Satisfaction (TCS)

Total customer satisfaction (TCS) means that all customers are satisfied before a job is considered successful. Many U.S. firms established some variant of TCS throughout their organization, and the concept then spread to service and government industries. In manufacturing, the introduction of TCS was important because it defined customers as both internal and external to the firm (in other words, at any process step, all downstream operations are customers). TCS recognizes that internal customers are critical for delivering a good product to external customers.

TCS was a significant development because it simplifies interaction in an organization by establishing the customer as the most important entity. It legitimizes methods to streamline work activities, thus making organizations more efficient (also known as being less bureaucratic).

6-Sigma Program

The 6-sigma quality improvement program was started in the late 1980s. The name comes from the statistical term used to describe variability in any process or product, with a higher sigma meaning that fewer defects are produced in the process (sigma will be explained later in Chapter 9). The program is comprehensive in promoting quality practices in manufacturing, including many of the tools used in this text.

The 6-sigma program is widely used by many firms, even though it uses an unconventional assumption in its statistical methods. The assumption is that all process means shift by 1.5 sigma (an unlikely event, but if this does occur then it should be reduced through improvement). This incorrect assumption makes it easier for any process to achieve a 6-sigma quality level by permitting more defects at a stated sigma level. In this 6-sigma program, a process that performs at 6-sigma can appear to have up to 3.4 defects per million parts made, versus a correct 6-sigma process which has only 2 defects per billion parts (over 1000 times more difficult to achieve).

Market-Driven Quality (MDQ)

The market-driven quality (MDQ) program was developed in the 1980s to apply quality techniques in all areas of a business (manufacturing, finance, purchasing, administrative services, etc.). It is an example of a program developed by one firm, but not widely accepted in the industry. It was a follow-up to another program (from the early 1980s) called Excellence Plus, which said that all work and product must conform to the specified requirement.

Total Productive Manufacturing (TPM)

Total productive manufacturing (TPM) is a refinement of the original Japanese TPM program, as discussed previously in the chapter. Both of these programs use the acronym "TPM," but the American program has broader objectives for manufacturing. It addresses equipment optimization, but also expands into other areas of manufacturing improvement, such as operator discipline and process improvement.

3.3 PROGRAMS BY CONSULTANTS

With the renewed interest in improving manufacturing in the 1980s came the growth of manufacturing consultants. The major programs developed or influenced in a substantial way by consultants are:

- *Total quality management (TQM)*
- *ISO 9000*
- *Malcolm Baldrige National Quality Award*

Total Quality Management (TQM)

Total quality management (TQM) is a quality management system applicable to diverse types of organizations including manufacturing, service, and government. It has gained widespread acceptance in all these businesses.

TQM is actively supported by consultants because its quality principles are based on a "top-down" management approach, striving to make quality filter down through the organization until arriving at the day-to-day operations. Its general philosophy and tools for managing quality are primarily suitable for management, especially executives.

The basic premise of TQM is that everyone in the organization strives to meet or exceed customers' expectations. It places customer satisfaction over short-term profits. The four key parts to TQM are people, continuous improvement, process, and the customer. A summary of each part is given in Table 3.2.[6]

ISO 9000

ISO stands for International Standards Organization, and ISO 9000 is a program designed to verify that a firm uses proper documentation, equipment calibration, and general control procedures for its processes. The ISO program was developed to demonstrate within and outside a firm that these controls are in place. A key part of the ISO program is an audit system to verify that documentation and procedures are controlled and traceable within a firm.

To attain ISO 9000 certification, a firm prepares its process documentation and control procedures to comply with guidelines published by the ISO organization.

---

[6] See Reference 5 in the bibliography.

Table 3.2 Four Parts to TQM

Key TQM Part	Description
People	People are trained on communication skills, interactive skills, and effective meeting skills. Additional training leads to improved teamwork and employee empowerment.
Continuous improvement	Employees gather data for intelligent decisions and ask "why" five times to get to the root cause. Improvement it erates with the "Plan, Do, Check, Action" cycle (known as the Deming cycle). Standardization is used to adopt suc cessful practices as the mode of operation.
Process	Employs a quality improvement process to define a step-by-step procedure to describe a problem in terms of cus tomer requirements. Solves problems by analyzing situa tion, developing solution, developing action plan, and im plementing plan followed by evaluation. TQM employs benchmarking to achieve "best-of-class" process.
Customer	The primary focus of TQM is the customer and customer satisfaction. Satisfied customers are essential to product and service differentiation over other firms.

Internal audits are a major part of the ISO review process. When the firm is ready, a team of consultants audits the firm to verify compliance, followed by new audits every several years. There are five different areas to the ISO 9000 certification, with most manufacturing firms seeking ISO 9001 or 9002, because these pertain to production processes.

The ISO 9000 program generates extensive consultant activity to help firms set up their documentation control systems and pass the initial and follow-up audits conducted by outside consultants. This effort is potentially costly in terms of money and resources for the firm; however, it may be worthwhile if the firm is small and must improve its control of procedures and documentation, and be capable of demonstrating this compliance to potential customers. Larger firms that already have control procedures in place will probably find that ISO 9000 serves mainly as a marketing tool.

Even though a firm has been successfully audited and accredited for ISO 9000, this does not mean that the audit procedures are correct, or that they are regularly adhered to on the manufacturing line. The audit only addresses the control procedures in place during the audit with no judgment of their value. An audit has obvious limitations in verifying daily compliance and can never replace the necessity of ongoing discipline and improvement.

Malcolm Baldrige National Quality Award

The Malcolm Baldrige National Quality Award began in the late 1980s to recognize companies that emphasize quality in their work process. The award is based primarily on self-audits in seven major categories related to management and quality, and has several award categories depending on the size of the firm and its particular industry. This award has brought more focus on quality in the workplace, with some firms using the audit structure as a foundation for improvement. The award, however, emphasizes the management of quality objectives in a firm, and has spawned an entire industry of consultants who willingly set up a quality management structure and assist companies to win the award.

3.4 EFFECTIVENESS OF IMPROVEMENT PROGRAMS

Considering that nearly all the manufacturing improvement programs started in the 1980s, it is astounding that so many were implemented in such a short period. It is a testimony to both the extent of the problems and the dynamism of U.S. manufacturing.

The challenge with such a broad range of improvement programs is to avoid confusion at the implementation level in manufacturing. If a firm attempts to understand its manufacturing problems and consistently uses a program to improve its process, then manufacturing personnel will understand the program's benefits and limitations. They become confident using the program in their daily work activities.

In many cases, however, improvement programs are not used properly. An improvement program typically starts out with good intentions, but after a period of time, its appeal is gone. Goals, statements, and measurement charts are published, with no concrete actions. A program may linger indefinitely, or be dropped in favor of a new one. Many times the right words are said, but the specific actions needed to change are not implemented at the process level.

To determine if a program will improve manufacturing, consider the following:

- What are the manufacturing process needs for improvement?
- What are the needs of manufacturing people to analyze and improve their process?

It is people with knowledge about manufacturing who can best answer these questions. This knowledge is acquired either by working in the process or by studying it. Ideally it is a combination of these two sources of information.

People are the fundamental source of improvement in manufacturing due to their unique ability to work in and study the process. Their level of commitment is critical for successful improvement. Manufacturing needs people to work together, pooling their technical and human resources as a team to overcome daily challenges for improvement. With committed, disciplined people, a company builds the manufacturing foundation for meeting the three basic criteria for success in a competitive market: lowest cost, highest quality, and shortest delivery.

SUMMARY

Important manufacturing improvement programs were developed or refined in Japan and then brought to the United States. These programs have been implemented in varying degrees at different firms. American companies have developed their unique form of improvement programs, with some programs accepted by other firms while others have remained company specific. There has also been a group of improvement programs fostered primarily by manufacturing consultants. These programs emphasize the management of the improvement process. The potential benefits from improvement programs should be judged from the standpoint of the specific manufacturing process needs, with special emphasis on the people working in manufacturing.

IMPORTANT TERMS

Zero defects
Quality circles
Continual improvement
Total quality control (TQC)
Statistical process control (SPC)
Just-in-time (JIT) manufacturing
Kanbans
5S manufacturing
CANDOS

Total preventive maintenance (TPM)
Total customer satisfaction (TCS)
6-sigma program
Market-driven quality (MDQ)
Total productive manufacturing (TPM)
Total quality management (TQM)
ISO 9000 (International Standards
  Organization)
Malcolm Baldrige National Quality Award

REVIEW QUESTIONS

1. Describe zero defects and how it is practiced in manufacturing.
2. What is a quality circle, and how could this benefit manufacturing?
3. Describe continual improvement.
4. What is TQC, and what are its five categories and their concepts?
5. How is SPC used to properly control a manufacturing process, and how is it used improperly?
6. What is JIT manufacturing? How are kanbans used?

7. What are the 5S and CANDOS programs, and how do they improve manufacturing?

8. What is total preventive maintenance and what is its approach to equipment maintenance?

9. Describe TCS and its major impact on manufacturing.

10. What is the 6-sigma quality program? How is this program used for manufacturing quality?

11. What is TQM? Describe its four parts for implementation.

12. What is the ISO 9000 program, and how does this affect manufacturing improvement?

13. What are the two considerations for assessing if an improvement program can benefit manufacturing?

4

✳ MANUFACTURING TEAMS

Just a few years since its inception into industry, team style of operations has become widely accepted. The concept of a team underscores the importance of individuals, each with their specific skills, coming from various technical and support areas to work together as a group to improve manufacturing. Teams recognize that no individual or work group has all the answers. Teams put responsibility and focus on the critical junction in a manufacturing firm: the process.

_____ **OBJECTIVES** _____

After studying the material in this chapter, you should be able to:

1. Describe a traditional manufacturing organization and discuss its benefits and risks.
2. Define *team* and explain the benefits of teams in manufacturing.
3. List the three types of manufacturing teams and discuss their pros and cons.
4. Describe the different technical members who participate in manufacturing teams.
5. List and discuss the attributes for an effective team.
6. Define *brainstorming* and describe the four steps required for its effectiveness.
7. Discuss the five skills needed by individuals for team participation.
8. Discuss seven leadership skills necessary for team leaders.
9. List and discuss the six reasons why teams fail.

4.1 TRADITIONAL ORGANIZATION

The **traditional organization** of manufacturing is based on a management hierarchy, where different manufacturing functions in the firm are represented by the managers and directors who report to the plant manager. The organizational functions are typically development, manufacturing engineering, manufacturing, quality, and support. The lowest level of management in manufacturing is the **supervisor,** who is responsible for all **operators** and **technicians** in a department. A **department** is usually focused on a specific operation in the process, with employees reporting to the supervisor. The supervisor answers to the higher levels of management.

The traditional manufacturing organization was common during the 1950s through the 1970s and is still widely used today. A sample organization chart of a traditional structure is shown in Figure 4.1.

This traditional management approach usually has several levels of **informal supervisors** between the manufacturing supervisor and the operator. These people go by various names, such as line tech, lead tech, or area coordinator. In essence, they carry out the wishes of the supervisor, ensuring that each operator or technician has work, knows the production priority, and so forth. In practice, supervisors spend most of their time dealing with management, while the informal supervisors (e.g., lead technicians) spend most of their time dealing with production issues. Management makes production decisions and these decisions are carried to the line operators through this hierarchy.

The traditional form of management is often inefficient because it adds layers of decision-making control between the operator and the manufacturing manager. It becomes difficult for any one person to make a day-to-day operational decision without management involvement. Management becomes overly focused on making decisions, thus forcing the organization to support its decisions, right or wrong, and creating the impression that change for improvement derives from management instead of the rightful source of the process. Traditional organization isolates management from the manufacturing process and stifles communication between people (both from within and between the different departments).

There are instances when this hierarchical management structure is efficient, with the supervisor involved in manufacturing issues with few coordinators needed to represent management. For instance, if the manufacturing situation requires

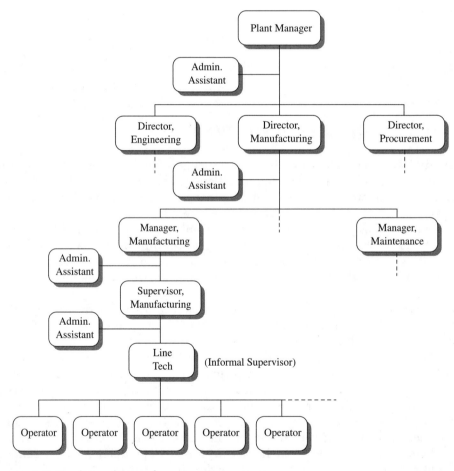

Figure 4.1 Traditional Manufacturing Organization

strict operator discipline with little need for individual decision making (such as building an older technology product with short delivery schedules), then the traditional approach may be effective by minimizing the number of people involved in decision making.

4.2 TEAMS

A **team** is a group of people working together with a common goal to meet routine production goals, complete projects, or solve problems. Teams have some decision-making responsibility within the limits of their members' capabilities. Teams start and develop for various reasons, such as two employees working together to solve an equipment problem, or a group of people responding to a management initiative to quickly introduce the production of a new product.

Manufacturing teams create the opportunity to mix **complementary technical skills** to improve the complex production process. The technical needs of manufacturing are so diverse that it is impossible for one individual to perform all tasks or to solve all problems ("two heads are better than one"). Because teams promote technical interaction, individuals become resourceful at providing or accepting information. Teams create technical self-sufficiency for day-to-day production operations. An interactive team organization is shown in Figure 4.2.

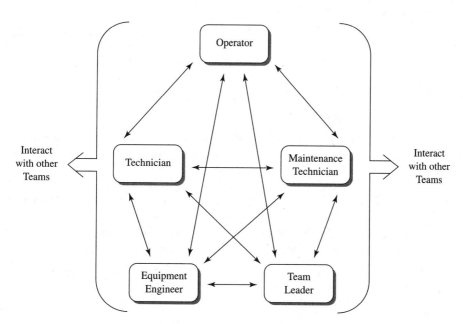

Figure 4.2 Self-Sufficient Manufacturing Team

Do not be misled by this team organization and think that team members can interact as they wish. The team has a clear structure with roles assigned to individuals based on their technical knowledge of the process and equipment. This group effort requires an understanding of each person's role and contribution. Because interaction among members does not pass through any one individual, a team is potentially an efficient organization for structured change.

A **manufacturing team** brings together individuals with different technical skills to determine the best course of action for solving production problems. The mixing of diverse technical skills in a structured team environment is healthy and strengthens discipline in manufacturing by using the power of many to overcome the limitations of one individual's insight.

4.2.1 Team Attitude

In the workplace, management may require your involvement in a team. You may then ask yourself, "Why can't I just do my job?" or "Why do I have to deal with so many other people?" Take a broader perspective of your operation and work, and develop a positive **team attitude.** Consider the consequences of letting work skills reduce us to detached "button pushers," with an "us" against "them" mentality, uninterested in the challenges needed to excel in manufacturing.

Each person in manufacturing is only one small part of the total process, with all the elements interacting as variables to produce the product. To improve the process and become more efficient, we have to place ourselves above the process elements while continually questioning our procedures and methods to thus increase value. We can realistically expect to achieve this organizationally complex task only through team interaction and support, with a positive team attitude.

4.2.2 Team Energy

Active team participation does not come about just from being on a team. It is only when we consciously choose to input human energy into a team that we accept the responsibility of active team membership. Team members with a positive attitude demonstrate this **team energy,** meaning that, any time we are on a team, we must take concrete steps to participate in the team in a positive manner for it to be successful.

Manufacturing teams have a special need for positive energy input from members because of the technical nature of manufacturing. Most people in manufacturing are overwhelmed by its complexity, learning to master their task at hand but never grasping the total process. To succeed in competitive manufacturing requires advanced technical, communication, and organizational skills that are most efficiently provided by the support of a team.

 BRAINSTORMING EXERCISE

Individual versus Team Perceptions

You will be given an exercise that compares team with individual judgments. The class forms into two groups: the members of one group work as individuals, while the other group forms into teams of three to four people. The individuals must work alone and should not look at or discuss the exercise with anyone else. A team can discuss this exercise only among its members.

Follow the instructions, making sure that no work is shared between individuals and groups. After completing this task, compare the results between individuals and teams (make a line graph on the board with the team answers in one color and the individuals in another color). Answer the following:

1. What differences did the class find between individual results and team results?
2. What can you say about the likelihood of finding the right answer for the teams versus the individuals?
3. Why would the teams come to a different result than the individuals?

4.3 TYPES OF MANUFACTURING TEAMS

The three different types of manufacturing teams are:

- *Informal team*
- *Formal team with a specific task*
- *Formal team with ongoing tasks*

Informal Team

An informal team is a group of employees working together to address a production need or solve a problem. They may not call or recognize themselves as a team; they may simply work together because they sense that a team effort is an effective way to solve a problem. For this type of team to form, employees need a conducive environment for working together as problem solvers. Employees work together as informal teams only if management supports this activity.

An informal team takes a minimal effort to start and maintain, because this team uses informal leadership. It potentially solves problems with minimal support and maximum efficiency. Informal teams tend to be adaptable, as a simplified organizational structure translates into ease of change.

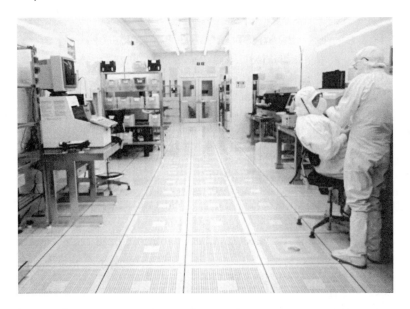

An informal team of two operators discuss their work.
Photo courtesy of Sematech Archives.

Informal teams do not work as well in a traditional manufacturing organization because interaction between people takes place within the hierarchical structure (see Figure 4.1). In a traditional management organization, managers often shun problems for fear of being blamed if it is difficult to correct. They want their employees to work only as directed to avoid having problems assigned to their group. This organization is not supportive of informal teams.

An informal team may be limited in the complexity of problems it can solve, as people from all work areas and technical skills are probably not represented on the team. Nevertheless, an informal team can be an effective way to resolve many of the day-to-day problems encountered in manufacturing.

✳ CASE STUDY ✳
Finding a Root Cause

A particular robot arm in a tool is causing wafers to jam and break approximately once a week. Each broken wafer costs at least $3000. In addition, the tool must be brought down during the run to remove the wafer and then requalify the tool, which is time consuming and costly.

The two operators from the day and night shift discussed this problem during shift changeover. They both noted that when the jam occurs, it always seems to happen with wafers near the end of a run (a run consists of twenty-five wafers in one cassette). The jam does not occur during every run, but when it does, it is consistently at the end of the run.

The day operator talked with maintenance, and a calibration check was done on the wafer autofeed mechanism that picks wafers out of the cassette and hands them off to the robot arm. A small accumulation error was found that caused some wafers at the end of runs to misalign in the autofeed mechanism. The tool was recalibrated, and in collaboration with the tool vendor, a setup gauge was designed and built to avoid this problem in the future.

Points to consider

1. Why did the two operators have insight about correcting this problem that might not have been available to one person or a larger team?
2. Is there a team in this case study? If so, who is on it?

A single-wafer robot arm in a semiconductor wafer fab.
Photo courtesy of Steve Martin.

Formal Team with a Specific Task

Formal teams with a specific task are temporary teams created by management to solve specific problems. They are sometimes referred to as a project team or a task force. Typical problems addressed by these teams include the following:

- Determine the cause of defects at a manufacturing operation
- Analyze why a tool continually breaks down
- Introduce a new process, product, tool, or procedure

Formal teams can be efficient for large problems, because they unite people from various technical areas. This cooperation speeds up problem resolution by blending both technical and nontechnical people to increase the problem-solving capability (different sources of ideas, different viewpoints, etc.). At the same time, large formal teams can become complex and restrict individual creativity.

Teams that exist for a limited time minimize the challenge of maintaining team energy over a long period. If the team is formed in response to a specific problem,

---

✳ CASE STUDY ✳
Maligned Teams

The manufacturing organization in a wafer fab is organized traditionally, with operators working under a line technician, who reports to the supervisor. The line technician is responsible for work assignments and daily problem solving. The operators have little ongoing interaction with the supervisor or management.

The company uses task force teams that enlist operators and technicians to work on general manufacturing problems, such as yield and equipment downtime. Operators are automatically on a task force team when a problem affects their work area, and most task forces have stayed in the same area since upper management created them 2 years ago. Each task force has a leader from outside of manufacturing who reports directly to a manufacturing manager.

Operators view the task force activities as an intrusion—just another task that gets in the way of their daily work. They joke among themselves about how the task force moves in to fix a problem when it is almost corrected and therefore takes the credit.

Points to consider

1. Why is the task force unable to mobilize people to improve the process?
2. Where do you think the task force leadership puts their focus: on the process or on what people think about the process?

then once the problem is resolved, the team is disbanded and people return to their regular work. The teams are focused on a particular problem, thus giving them purpose for temporary existence.

Formal Team with Ongoing Tasks

Some modern manufacturing firms are organized around formal teams with on going tasks that exist indefinitely. Employees work in this formal team organization, with a supervisor who may be called by various names, including team leader or team facilitator. The different technical disciplines in the team will vary depending on how the team is organized. For instance, it may be a self-sufficient team that includes all technical people to support a process, including operators, technicians, engineers, maintenance, and other areas (cross-functional team).

In a manufacturing firm with ongoing formal teams, the team has replaced the traditional organizational hierarchy. The operator and technician work within the team, receiving direction and getting technical support from other team members and the team leader. Day-to-day decisions are made at the level of the manufacturing process, including the identification and resolution of problems.

A formal team does not replace the problem-solving role of an operator, but potentially provides a means to solve many problems at the team level without involving external supervisors. A team workplace organization requires less intervention by off-line people removed from the manufacturing process.

On the other hand, an ongoing formal team requires substantial human energy to sustain positive team action. It is easy for an ongoing formal team to deteriorate for lack of sufficient energy and to remain as a team in name only, while actually representing a traditional organization.

4.4   TEAM MEMBERS

Manufacturing team members should come from all technical areas of production and various support groups. The team member skills for a particular team vary depending on how the team is formed and organized. A **product-based team** is formed around a particular product and is self-sufficient for all skills needed to produce the product. On the other hand, an **operation-based team** (also known as a function-based team) is formed around a specific operation, or function, and is self-sufficient for that operation.

Ideally, team members have complementary skills, each contributing in their unique manner toward solving manufacturing problems. For instance, a supervisor will address problems that require management intervention, whereas an operator investigates problems with support tools used at the workplace.

The primary skill areas for team members are:

- *Operator*
- *Equipment and maintenance technician*

- *Process technician*
- *Supervisor*
- *Equipment engineer*
- *Process engineer*
- *Support personnel*

Operator

The operator is the person primarily responsible for operating the equipment in the factory where the product is built. All expertise to run the operation resides in the operator, including knowledge of the procedures, product requirements, quality standards, and equipment operation. Operators undergo training to learn their tasks and are typically certified to operate certain equipment. Their competence in product, equipment, and process requirements is critical for achieving standardization of work procedures and repeatability of the process.

The operator has different titles, including manufacturing technician, specialist, associate, and wafer fab technician. In the last several years, the trend in some firms has been to eliminate the term *operator* and replace it with *technician* or *specialist*. As long as this new title actually represents a technical skill level, then it is appropriate. This book makes a distinction between the two skill levels, while recognizing that the terms could describe the same person.

Equipment and Maintenance Technician

Equipment and maintenance technicians are responsible for troubleshooting and maintaining the equipment used in the factory. In many cases, technicians supervise or provide equipment-related support to operators. The equipment technician and operator can also be the same person or different people on the same team. This integration of roles is an important part of the total productive maintenance (TPM) improvement program.

Process Technician

A process technician is responsible for ensuring that products are properly manufactured on schedule at their assigned operations. Process technicians provide process-related support to engineers, operators, and equipment technicians. The process technician may also be referred to as the line technician and be assigned broad duties by the supervisor.

Supervisor

A supervisor is a manager responsible for a group of operators, technicians, and possibly engineers in manufacturing. Supervisors ensure that products are properly built on schedule; they are also responsible for personnel issues. Because supervisors interact with employees, a large share of their time is spent ensuring that employees fulfill commitments made by management. Supervisors have various titles, including first-line manager or operations facilitator.

Equipment Engineer

The equipment engineer is responsible for defining and implementing the equipment used to manufacture product. This could include equipment design and build, production capacity assessment, troubleshooting, and ongoing support of specific equipment used in manufacturing.

Process Engineer

The process engineer defines the necessary settings for equipment and process variables and communicates these specifications to manufacturing through written documentation, personnel training, and ongoing process improvement. The process engineer should be instrumental in defining the critical variables and settings for the process.

✳ CASE STUDY ✳
Material Supplier

Two companies supply silicon wafers to a semiconductor wafer fab. Wafer quality requirements include criteria for flatness, wafer dimensions, and cleanliness. An edge grind is put on the outer edge of the wafer to create a smooth edge surface (which reduces defect growth in the silicon crystal).

Semiconductor fabrication requires stringent cleanliness control to avoid contamination on the wafer. Wafer fabs are designed around cleanroom workbays with strict contamination control of humans, equipment, and materials.

While processing incoming wafers, operators notice that wafers from one of the two companies have excessive particles on them. Investigation with a high-powered microscope finds that the edge grind process done at the supplier appears rough, and could be the source of the particles. The operators inform their supervisor of the particles, but because the supply pipeline is already full of these contaminated wafers, they are forced to clean and use them. To the extent possible, they use wafers from the other supplier that have a smooth edge grind and no visible particles.

Points to consider

1. Do the quality practices at the supplier have a direct effect on the quality output of the wafer fab?
2. Are the wafer suppliers a part of the production process?
3. Should the wafer suppliers be considered part of a production team?

Support Personnel

Manufacturing requires extensive support personnel, such as people working in shipping, materials support, and accounting. These people are sometimes referred to as indirect, because they do not work directly in the process; however, their skills are critical for the process to function. Other support people are the material suppliers who supply raw and partially finished materials, and equipment field representatives who support their equipment in production.

4.5 EFFECTIVE TEAM ATTRIBUTES

Team attributes are characteristics that can be used to describe how team members work together. Following are the desirable attributes of an effective team:

- *Open communication*
- *Defined goals with measurements*
- *Complementary skills*
- *Clear roles and responsibilities*
- *Acceptance of team leader*
- *Brainstorming*
- *Clear decisions*
- *Commitment*
- *Interaction with other teams*

Open Communication

An effective team has open communication channels among team members and with other teams throughout the process, and thus permits team members to express their ideas and concerns without fear of retribution. The most fundamental aspect of open communication is **respect** for other team members, including fair recognition of work and accomplishments by the team members.

Open communication must also provide a means to resolve problems between team members. Ideally the problem resolution is achieved by talking with one another and obtaining support from the team leader, if necessary. Communication is often more efficient if the team is a **small group.** This simplifies interaction between members and avoids excessive coordination. People encounter more communication problems when forced to interact with too many different people to get their jobs done.

A small group may appear unachievable for complex manufacturing problems, where the team needs to interact with multiple technical areas. Even in this situation, however, a team can form smaller working subgroups to efficiently address particular problems. These smaller groups can function as informal teams within the overall formal team structure.

Defined Goals with Measurements

Effective teams **establish goals** that all the team members understand, and develop measurement criteria to ensure that teams meet these goals. Teams can then develop a **team charter** that states their objectives and defines the roles of the different members. The team charter should also state general team ground rules for how members interact to accomplish the goals.

The team needs general and specific goals. For instance, if a team wants to improve a particular operation, then general goals for improvement are outlined (e.g., less equipment downtime with an increase in the number of parts processed). The team then must address specific goals, such as the actual problems that cause the tools to break down and the specific repair actions needed to improve the downtime. They could compare the performance of the tool to supplier specifications or performance data from similar tools. Measurements are efficient when the team collects data and uses that information in their decision-making process.

Complementary Skills

The way to address technical problems in manufacturing is to have *self-sufficient teams with complementary skills*. Each individual brings technical expertise to the team, which increases the probability of finding solutions. Individual backgrounds and skills should be discussed with all team members during the early formation of the team so that people understand the technical expertise that exists within the team.

Clear Roles and Responsibilities

Once the skills of the team members are identified, then it is effective to assign tasks to those who can most easily address appropriate issues. This becomes clear for the team when skills are balanced against the range of problems confronted by the team.

Team members are accountable for their own personal actions and for meeting team commitments. Individual responsibilities can contribute toward both the overall team function and specific problems undertaken by the team. An example of a team responsibility is documentation, such as recording the minutes of the team meeting. A responsibility related to a specific problem could be properly documenting corrective action in an equipment log.

The team must hold all members mutually accountable for their effort on the team, and how their activity supports solving manufacturing problems. If the team deteriorates into a "club" activity with people complaining or talking about what they would like (e.g., a pay raise or more break time), then the team is a waste and should be stopped. Teams must stay focused on their goals.

Acceptance of Team Leader

The team leader guides the group. To be effective, the team leader should acknowledge the special role team leadership requires, while team members strive to accept the leader. In many cases, the team leader is appointed by management,

or is automatically assumed to be the manufacturing supervisor. In other cases, team members take turns at team leadership, which could be a positive idea depending on the different members in the team.

The team leader can be a facilitator, or the facilitator may be a different person. The facilitator guides the team to stay focused and makes team-based decisions. The facilitator can come from within the team, or lead from outside the team by meeting with the group when necessary.

Sometimes the team leader or facilitator (depending on which is used) is elected by the team and is not a manager. The team may also find that sometimes the best team leader is naturally acknowledged with no formal election.

Brainstorming

Brainstorming is an idea-generation technique based on an open exchange of team ideas to find the optimum solution to a problem. If done properly, brainstorming unleashes the power of the individual by feeding ideas off of each team member. The four brainstorming stages or guidelines are:

1. *Organize the session.* Select a team leader that is acceptable to everyone. Clarify the ground rules for the brainstorming session. Appoint someone as the official recorder of the team, so all ideas are made visible to the team. Use a marker board, flip chart, or cards to exchange ideas.

2. *Describe the problem.* All team members must understand the problem. It can be restated several ways to ensure that all aspects are clear.

3. *Generate ideas with no criticism.* Rotate between team members seeking ideas, letting the ideas flow freely to generate as much input as possible from everyone. This is the hardest part because some people are prone to giving criticism. No one should feel inhibited. Everyone has an opportunity to contribute, but if a team member does not have an idea, then he or she can "pass" until the next round. Equality between members and sincere contribution are the foundations of brainstorming, because any idea may trigger someone's creativity toward finding the optimum solution.

4. *Evaluate all ideas.* Judge ideas for practicality and effectiveness. It can help to group similar ideas, and then try to reduce these ideas to a single concept. Focus on ideas that could be realistically implemented by the team. When using an evaluation approach, each team member ranks the ideas using the following scale:

    0 This idea will not help solve the problem.
    1 There is a chance this idea will be useful.
    2 This idea should be of some help.
    3 This idea will definitely help in solving the problem.

After all members have completed their personal ranking, add the totals for each idea. The ideas with the highest totals are selected as most likely to solve the problem.

BRAINSTORMING EXERCISE

Problem Solving

Use the four brainstorming guidelines to find as many solutions as possible to
the following:

a. How many different ways we can use a paper clip

b. Ways to keep students from dropping out of college

c. How to conserve more water in residences and businesses

Clear Decisions

Teams need clear decisions to resolve manufacturing problems. It is important to discuss the decision-making process within the team, and have it agreed upon by all
team members early in the formation of the team. This groundwork permits people
to understand how decisions are made and how they can input suggestions. The decisions should be agreed upon within and outside the team, and be clear so that affected people understand them, to help minimize the situation of one person making
decisions without team input.

Manufacturing team decisions must resolve production problems. If the team is
working on problems that do not address improvement issues, then the team is not
properly focused.

Commitment

Committed individuals are indispensable for positive team activity. The team effort is
a microcosm of the company's effort, and each individual should sense the personal responsibility. Ideally, problems arising from personality conflicts, misunderstandings,
and egos are dealt with at the team level, preferably with effective team leadership.

Interaction with Other Teams

Manufacturing teams interact with other teams to understand their mutual problems and how they can improve the process. Teams must avoid a parochial view of
"us" versus "them." Team leadership plays a major role in assisting interaction between the different teams.

4.6 INDIVIDUAL SKILLS FOR EFFECTIVE TEAMS

Teams succeed because individuals place the team needs above their own needs. Because teams are composed of individuals, the effectiveness of a team is limited by the
effectiveness of each individual. One person can damage a team beyond repair. On

the other hand, the power of the team can go beyond the sum of each member's contribution, thus enhancing the concept of *team spirit.*

Individual skills for effective teams are:

- *Listen and talk*
- *Work for the team's success*
- *Avoid criticizing team members*
- *Contribute at the individual and team levels*
- *Be open-minded*

Listen and Talk

As a team member, you interact and communicate with other team members within and outside your team, which means you contribute input to the team or accept input from someone else on the team. Make a sincere effort to listen to other team members when they are talking, and try to understand their proposals. If you have an idea, offer it to the team for their review. Each team member must make an effort for positive communication between all individuals.

Work for the Team's Success

We each have a sense of our own individuality, but have to work together to create a sense of our team's identity. Once you are part of a team, place the team's goals above your individual needs. If a properly developed team can achieve this higher sense of purpose for each team member, then it creates real team spirit, with a capability far greater than that achievable by any one individual.

Avoid Criticizing Team Members

Discuss problems, discuss alternative scenarios, express technical concerns, or make suggestions to reach team consensus. All of this is permissible and desirable to achieve the optimum team solution. However, teamwork does not permit criticism of other team members. Avoid situations that pit one individual against another through criticism—they only weaken the team by creating distrust. Restricting criticism instills confidence in team members that they can convey their ideas without fear of public ridicule.

We have seen how brainstorming restricts criticism of ideas during the idea-generation stage. It is also important to avoid criticism during one-on-one interactions between team members. Each team member should develop a sense of respect for the other members, but this may be difficult depending on the individual members and their past history. If personal respect is a problem, then find a common ground for respect of one another. If this is impossible, recognize the problem and change team members.

Contribute at the Individual and Team Levels

An effective team member works with individuals or groups in different types of settings and in a wide range of activities, such as operating the equipment, determining how to rearrange the workplace in the most efficient manner, documenting problems from one shift to another, or attending team meetings. Having the ability to interact positively at both the individual and group levels is important for success as a team member.

Be Open-Minded

Sometimes people have been working at an operation and already have set ideas about how it should be run and what the problems are. Insight and experience are important, but avoid the NIH (Not-Invented-Here) syndrome. Having a closed mind to other ways will hurt the team's performance.

4.7 LEADERSHIP SKILLS FOR TEAMS

The team leader provides guidance to the individual and group. The team leader's key contribution is the ability to take a global approach, both within the team and when dealing with other teams, while maintaining a local perspective at the level of the individual. An effective team leader must be able to work at both levels simultaneously, even while serving as a contributing team member.

When an operator or technician participates in a team, there may be a tendency to view the team leader as someone above the team. An effective team leader dispels this notion. The team leader has a guidance role in the team, but it is actually no more important than the role of each team member to contribute personal and technical skills.

Furthermore, team members should be willing to contribute as a team leader. The case can be made that any one person could be either a good leader or a poor leader. It is surprising how a person who no one thought was capable of leadership can become an effective team leader. Leadership skills are acquired by exposure to various types of life activities, in such a manner that we may be unaware we possess leadership skills until required to use them.

Skills for effective team leadership are:

- *Be a coach*
- *Treat people fairly*
- *Care about others*
- *Promote team-wide participation*
- *Give credit*
- *Keep the team focused*
- *Develop intra-team relationships*

Be a Coach

Team leaders persuade or urge other people to achieve their individual and team potential. The team leader provides support to these people, assisting them to grow in their responsibility for tasks and problem solving. In many ways, an effective coach is transparent, providing subtle support when needed, and knowing when to back off and permit people to make their own decisions. In this environment, team members accept responsibility for their decisions, yet appreciate guidance from the coach.

Treat People Fairly

Team leaders create team division when they do not treat all people fairly. Sometimes team members might not agree with the team leader's actions, but each team member should always feel that the leader has acted fairly.

Care About Others

Teams are made of individuals who have feelings about their work and their role. The team leader should be sensitive to individuals in the group.

Promote Team-Wide Participation

A team can never be a fully functional team if the individuals do not participate to the best of their ability. The team leader must assist each individual to contribute personal and technical knowledge for the good of the team.

Give Credit

The greatest attribute of an effective team leader is giving proper credit to team members who accomplish a task. The credit can be given at the team level or could include recognition by a larger group. Giving proper credit is the most basic form of

---

 BRAINSTORMING EXERCISE

Team Leadership

You are the team leader of a manufacturing team in the photolithography area of manufacturing. A new, state-of-the-art tool is coming into the photo area in 3 months, and a team member must be chosen to receive training as the key operator for this tool. You believe a particular operator, Josephine, would be the best operator to receive this training due to her technical ability and positive attitude with other employees. Another team member, Sally, does average work but is always vying for promotion. Sally has made it known to the team that she wants to be trained as the key operator for the new tool because she has worked at her position the longest.

What do you do as team leader?

respect by a leader to the team members who are doing the work, and goes a long way toward establishing leadership credibility.

It is all too easy for a team leader to improperly take credit for work that is actually the team's accomplishment. People who work directly in a process know who contributes to get the job done. Team leaders from an outside area (e.g., a facilitator who comes only to team meetings) easily fall into the trap of assuming they are doing the work. Worse, they cultivate this image among people outside the team. Because they are not directly involved in the ongoing work of the people in the process, they lose sight of reality.

If a team leader takes undue credit for solving a problem, everyone doing the work becomes aware of that fact. In most cases, it is unreasonable to expect an operator or technician to resolve this problem because it involves someone perceived as above him or her in the organization. The team no longer functions as a team while the leader continues to exploit his or her position of power to take more credit from the group. Improper credit or lack of recognition leads to contention and low morale among team members.

Keep the Team Focused

Teams are ineffective if their efforts are not focused on accomplishing specific goals. Although the team is responsible for setting goals and measuring themselves, it is the team leader who takes an overall view of the team to ensure that individual actions are consistent with focused team goals. If the team's actions are unfocused and manipulated by the leader for self-gain, the ensuing politics will harm the team.

Develop Intra-team Relationships

An effective team leader interacts with other teams to ensure that all teams are working toward a common objective. This assessment requires stepping aside periodically to look at the bigger picture and ensure that the team fits in the total process activity. The responsibility of the team leader requires adept personal skills to separate and understand the larger process while working with team members on local problems.

4.8 PITFALLS OF TEAMS

Teams may appear to be the solution to every problem in manufacturing. Yet there are **team pitfalls** that potentially create problems in manufacturing. We need to understand these drawbacks and be able to confront them as they arise. The pitfalls of teams are:

- *Team members lose enthusiasm*
- *Team used for someone's advantage*
- *Inadequate positive contributions*

- *Team lacks focus*
- *Team used for day-to-day management*
- *Team for all situations*

Team Members Lose Enthusiasm

The tendency for any ignored process is toward decay and randomness. A team endeavor is a process and is susceptible to this natural law. If no effort is put into a team, it will lose enthusiasm and become nonproductive. It takes energy, specifically human energy, to form and maintain a team. This energy is easily lost if each individual does not make a sincere effort for the team to succeed.

We maintain team enthusiasm by instilling a sense of freshness in the team members. If a team solves a problem that was previously thought unsolvable, then members sense excitement.

Teams need a balance between the individual and the team activities. Allow individuals time to do their work, to think through the problems, and reflect on solutions proposed by the team. Avoid forcing solutions on team members.

Team Used for Someone's Advantage

Unfortunately, there will always be individuals who place their personal gain above that of the group, thus creating group tension and keeping the team from functioning properly. When this occurs, the culprit is often the "leader" who has more interest in his or her own career path than in the success of the team or firm. Excessive "politics" becomes an issue, then, when team members' work is used to benefit the leaders rather than to improve production. This degenerate condition can occur at the group level, and in some cases, throughout the entire company.

Constructive team members who place group gain above personal gain must work together, hopefully under the guidance of the team leader, to try to neutralize negative activity. A possible approach is demonstrating how each individual benefits by working as a team. Some situations have no easy solution, which may indicate that it could be more effective to disband the team.

Inadequate Positive Contributions

Some individuals choose not to make positive team contributions. They may feel coerced by the team, disagree with its goals, or sense the team is contrived. These individuals believe working alone at their workplace is a more effective way to solve problems.

A team sometimes threatens individuals because they perceive a loss of their power in the organization. At work, we typically establish relationships between coworkers, creating "comfort zones" where people become comfortable with their roles and responsibilities vis-à-vis their coworkers. Forming a team can challenge this relationship.

Some people fear the formation of a team because it potentially uncovers previously hidden mistakes. The team goal is to improve, not to lay blame or point fingers. Positive team activity can set an example by showing how the team recognizes that mistakes are made. Only by addressing the problems can the team advance and minimize future mistakes.

Note that at times an individual's misgivings are correct, which serves as a sign to the team that there is a problem. The team leader must objectively try to comprehend these feelings, and then compare them with the team's actions to see the reason for the problem. This is team leadership.

Setting an example is an effective way to involve noncontributing team members in the team. If the team's activities are getting results, show this to the unwilling team members in a nonthreatening manner. Ask their opinion about the results, or get their input on how to proceed. In a constructive manner, let the team's positive results serve as a stepping stone for involvement by the noncontributing team member. People like to be part of successful activities, and the team can use this to its benefit. This is all in the spirit of getting everyone involved—some sooner and others later.

 EXAMPLE 4.1   RESISTANCE TO CHANGE

There is a recurring damage problem to a photomask reticle (the master pattern) used in the photo workbay. Several operators are opposed to analyzing the problem. They believe the problem is caused by the photomask supplier, and there is nothing that can be done about it except complain about the supplier's quality standards.

What are possible approaches for achieving a positive effort to solve this problem?

Solutions

1. Talk to the members and find out why they believe it is a supplier problem. Explore possibilities that the source of the problem could occur both at the supplier and in house. Brainstorm ways to immediately contain the problem and then solve it.

2. Use "What if?" to increase problem perspectives. For instance, "What if the problem was due to handling damage?" or "What if the supplier was informed we have this problem?"

3. Discuss how to objectively assess where the problem is occurring. A temporary inspection could be done when the photomask reticles are received to determine if the damage is coming from the supplier. If reticles are good when received, then the team knows it is an in-house problem. Attempt to get participation from all team members to analyze the source of this problem.

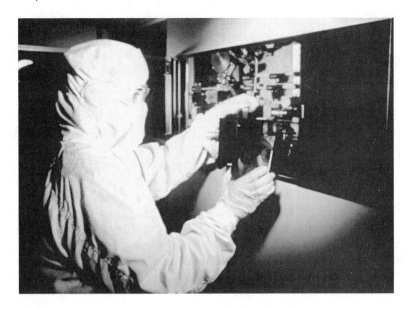

Wafer fab operator examines a photomask reticle for defects.
Photo courtesy of Sematech Archives.

Team Lacks Focus

The main reason for lack of team focus is inadequate leadership. Individuals contribute toward the team's problem-solving goals, while the team leader guides integration of the individual actions into the team effort. A poorly focused team makes little progress toward solving problems and creates demoralized team members.

An unfocused team is detrimental because problems remain unsolved and can actually grow in severity. The fundamental reason for the team's existence is to solve problems, which is not accomplished on an unfocused team. In this situation, the team may need an objective viewpoint from outside the team to provide direction, or it may be necessary to seek new team leadership.

Team Used for Day-To-Day Management

It is acceptable for managers to use a team as the organization for managing their employees. Unless the concept of a participatory team is maintained, however, this situation can easily change into traditional management. For example, a manager may have several teams, each with its respective team leaders. The manager relays instructions to and from the teams through the leaders with no input from the team members, creating contention between the team members and management's desire for employees to follow their direction. The operators are part of a "team," but managed in the style corresponding to the needs of a manager. This situation is hypocritical and leads to dissatisfied employees.

The risks of a traditional management organization within a team structure are (1) loss of creativity, (2) lack of individual spontaneity or participation, and (3) employee confusion and demoralization. If management chooses to organize manufacturing around teams, they must support team activities by respecting the teams and their role in manufacturing. Management's goal is to develop team responsibility while avoiding team control. Otherwise there is little reason to have teams.

Team for All Situations

Imagine yourself a private in the military, and the sergeant crawls in the trench and informs the platoon that there will be a team meeting to decide if the platoon should attack the hill tomorrow morning. A fellow private, in the spirit of team participation, offers an opinion that the attack should be delayed several days so the platoon could brainstorm the situation. In this case, a team approach may not be the best way to make a decision.

There are manufacturing situations when team organization could create problems. For instance, if a new and unstable manufacturing process has an immediate need to increase production, then the volume ramp-up may be more effective with experienced technical leaders setting clear direction for production.

Another situation when teams may not be optimum is a new start-up firm in an established business. The operators and technicians have a high skill level. In this case, strict operator discipline creates an atmosphere of precision and control that attracts potential customers. A participatory team approach may not be the most effective way to gain customers in the short term; however, a team is the best long-term organization for implementing positive improvement actions because an established process typically has many sources of inefficiency and waste.

SUMMARY

The traditional organization of manufacturing is based on management who relays instructions to workers through supervisors and informal supervisors. Decision-making authority rests with management. An alternative manufacturing organization is through teams, permitting day-to-day decisions to be made at the team level in the process. Effective teams require a positive attitude and constructive energy from each member.

There are informal and formal manufacturing teams, with all manufacturing work areas represented. Effective teams have certain attributes that contribute to their success. Individual and leadership skills are essential for a team to function properly. Teams also have pitfalls that, if not properly addressed, can impair their success in manufacturing.

IMPORTANT TERMS

Traditional organization	Product-based team
Supervisor	Operation-based team
Operator	Team attributes
Technician	Respect
Department	Small group
Informal supervisors	Establish goals
Teams	Team charter
Complementary technical skills	Team leader
Manufacturing team	Brainstorming
Team attitude	Coach
Team energy	Give proper credit
Informal team	Team pitfalls
Formal team with a specific task	Team enthusiasm
Formal team with ongoing tasks	

REVIEW QUESTIONS

1. Describe the traditional organizational structure for manufacturing. What are some pros and cons of this organization?
2. What is a manufacturing team? Describe the potential benefits from teams.
3. List and describe the three types of manufacturing teams.
4. List the technical members on a manufacturing team.
5. List and discuss the attributes of an effective team.
6. What is brainstorming and what are the four stages or guidelines?
7. List and discuss the skills necessary for individuals on a manufacturing team.
8. What skills are necessary for team leadership?
9. List and discuss the pitfalls of manufacturing teams.

EXERCISES

Traditional Organization

1. You have been working for 2 years as a certified operator at a wafer fab workstation. You know the requirements for building good product and work under a lead technician who respects your knowledge and work ability. The lead technician assigns you overtime, weekend work, and special tasks from time to time. Everyone in the department works for a supervisor, who delegates most of the manufactur-

ing work supervision to the lead technician. When you have a problem and need assistance, you always go to the lead technician first.

Recently, the lead technician has been assigning another operator to work at your workstation during peak workload periods. This operator has no training and is not certified. You know that defective parts are produced and passed on when the uncertified operator is working. You discussed this with the lead technician, who said it must be done to "make the numbers," and it was not your responsibility. You would like to talk to the supervisor about the other operator's lack of training, but feel awkward going around the lead technician.

A. If you were the operator, what would you do in this situation?
B. If there are defective parts and this becomes an issue as to why they occurred, who do you think will be blamed?
C. Can you recommend a better way to organize the department to avoid this type of problem?

Teams

2. Reconsider Exercise 3 in Chapter 1. Explain how you recommend that the operator who noticed the lack of adherence to procedure should address the problem if the department is organized as:

A. a traditional organization, with a line technician who gives them both direction, and a supervisor.
B. a team organization, where they interact and make team decisions with a team facilitator.

Brainstorming

3. Consider the following brainstorming scenario.[1] What can you find wrong?

Big Boss charged into the department meeting. He had a new idea. "Today we are going to have a BRAINSTORMING SESSION! We are going to solve our Big Problem! Somebody give me an idea!" Always loyal, Right-Hand gave Big Boss an idea: "We should do X. X has always worked for us before." Big Boss looked upset. "X won't work. That's why we're having this session."

Rising Star saw a chance. "Let's try Y. The president always likes it when we try Y." Big Boss rolled his eyes. Rising Star was always trying to please the president. Everyone noticed that Big Boss was not pleased, and became quiet.

Then Hare Brain perked up. "We could try Z. Z might work. We will never know if we don't try." That was too much. Everyone laughed. "Z could not possibly work," they told Hare Brain.

Big Boss was frustrated. "Brainstorming is for the birds!" he yelled as he left the meeting, carrying his Big Problem on his back.

---

[1]See Reference 4 in the bibliography.

Pitfalls of Teams

4. You work the third shift in a wafer fab, from 11:00 P.M. to 7:30 A.M. The company is organized in a traditional structure, and has formal teams to work on diverse manufacturing problems such as yield improvement or equipment breakdown. The teams make recommendations to supervisors, who then implement changes in their department. The weekly meetings start at 7:30 A.M. (after the end of the night shift) and last 1 hour. The meetings are voluntary and operators are paid overtime to attend.

   The company enters a severe business slump, and cancels all overtime. You notice that attendance at the team meetings drops by over 75 percent.

   A. Why are people attending these meetings? What observations can you make about team enthusiasm?

   B. What suggestions do you have for improving how teams function in this company?

5

✳ THE PROCESS

A manufacturing process is the repetitive sequence of process steps that add value to build a functional product. Stated another way, the process is how products are made in manufacturing.

Can we really say that a manufacturing process is merely how we make something? If you enter a factory, and can stand in the middle of it, there is more than what meets the eye. You can actually feel something. That "something" is the human energy and work that has created the process, giving a sense of purpose much larger than the work in front of you.

Given the complexity of a manufacturing process, to study the entire subject matter in a chapter would be difficult (we would probably need a thick book to do it justice). To simplify our task, the objective is to study the topics directly related to meeting our three goals for competitive manufacturing: lowest cost, highest quality, and shortest delivery.

OBJECTIVES

After studying the material in this chapter, you should be able to:

1. When given a key manufacturing process term, discuss it with respect to competitive manufacturing.
2. Explain how excessive WIP can cover up manufacturing problems.
3. Draw a chart showing how a process improves from high defects and inspection to low defects and inspection with minimal waste.
4. Calculate and interpret yield and total process yield, given the input of parts and the number of good parts out.
5. Calculate and interpret cycle time and throughput, given the production quantity and time.
6. Understand how series versus parallel and single-piece processing improves cycle time.
7. Discuss the holistic process view, why it is necessary, and how to achieve it.

5.1 PROCESS TERMINOLOGY

High-volume manufacturing has extensive terminology to describe what occurs in the process. These terms are used to communicate technically with other manufacturing people. Our approach for studying the process is to start at the center of the process, the workplace, and expand outward to encompass the critical process aspects that affect competitive manufacturing. The key manufacturing process terms are:

- *Workstation*
- *Work in process (WIP)*
- *Inspection*
- *Yield*
- *Throughput*
- *Cycle time*
- *Process flow*
- *Specifications*
- *Training*
- *Process control*

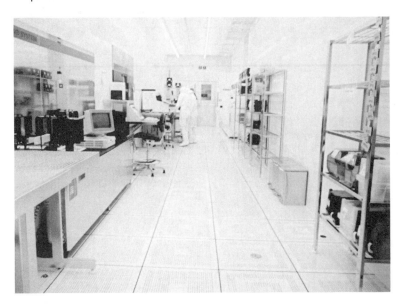

Operator at a workstation in a workbay.
Photo courtesy of Sematech Archives.

Workstation

The workstation is the actual location on the manufacturing floor where the operator and technician work. There are different types of workstations. A work cell integrates multiple tools at a workstation with one operator controlling the equipment. In a semiconductor wafer fab, most work is done in workbays to reduce contamination and simplify the delivery of process materials needed to manufacture the product.

The amount of product processed per unit time through equipment at a workstation is **equipment capacity.** For wafer fabs, equipment capacity is usually specified as *wafers per hour,* or *WPH*. It is good when equipment is run just below capacity (90 to 95 percent of capacity), with a 5 to 10 percent buffer for emergencies. The more stable the equipment, the less buffer required.

The **setup time** is the time it takes to change over equipment to run different products, and is time unavailable for production. Setup time includes activities such as changing fixtures, downloading computer software, and setting different equipment parameters. How operators conduct their work at the workstation has a big impact on setup time. Shorter setup time increases equipment capacity and flexibility in multiple-product lines because the equipment is more available for production.

Prior to the 1980s, equipment setup time was ignored, as firms processed large lot sizes to avoid frequent tool changeover. This was acceptable when firms had limited types of products in the capacity production era of the 1950s, but it has a negative effect on manufacturing efficiency in a competitive environment.

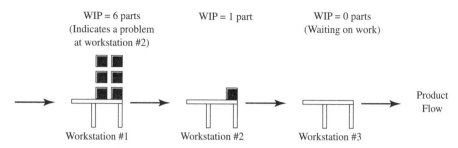

Figure 5.1 Excessive WIP Indicating a Problem

Work in Process

Work in process (WIP) represents the product at the various steps in the manufacturing line, from incoming components, through build and assembly, and up to shipping. The actual sequence of steps that WIP follows through the process is defined for different part numbers based on processing requirements, and is controlled in manufacturing by a **floor control system** using software or manufacturing floor documentation. This system controls when WIP enters and leaves the workstation, plus logs any other pertinent information (e.g., the number of reject product). Managing WIP flow through process operations can consume substantial operator time.

Excessive WIP at an operation indicates process inefficiency, and is an important indicator to the team that there is a problem. For instance, WIP buildup occurs when an operation stops due to a problem (such as a broken machine), yet upstream processes continue to build. In this case, *WIP accumulates in front of the operation, highlighting the problem.* This is shown in Figure 5.1, where Workstation 2 is not able to produce at the same rate as Workstation 1, thus causing excessive WIP at Workstation 1, with insufficient WIP at Workstation 3. To attain smooth product flow, the reason for the line imbalance at Workstation 2 must be found and corrected. When corrected, the WIP will reduce in the process. Note also in Figure 5.1 that Workstation 2 is the only workstation with the correct amount of WIP, yet it is the bottleneck!

WIP is also related to the number of parts grouped together for processing, known as **lot** or **batch processing.** In wafer fabs, wafers are processed in batches of typically twenty-four or twenty-five wafers, although many integrated tools process wafers individually. If the lot size increases, flow in the process becomes discontinuous because some parts are always waiting at operations for the remaining parts in the lot. This leads to increased WIP, disrupts process flow, and is contrary to the goal of reduced delivery time.

WIP in a manufacturing process can be viewed as water flowing in a river. A deep portion of the river has excessive WIP, covering dangerous rocks that represent problems. To have efficient process flow, we need to lower the WIP and correct the problems (lower the water level and remove the rocks). This example is shown in Figure 5.2.

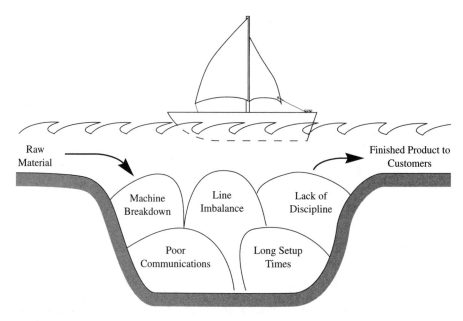

Figure 5.2 Excessive WIP Covers Process Problems
Redrawn from "The New Manufacturing Challenge," K. Suzaki, 1987.

Inspection

Disciplined work requires all people to be constantly aware of the product and how it conforms to the requirements. Inspection is the effort to verify that the product conforms to the requirements in the specification. It is done several ways, including visual inspections with instruments such as microscopes or with semiautomated and automated inspection systems such as an ellipsometer to measure wafer-film thickness. Inspection requires defined quality criteria that are understood by all team members.

Inspection done at the workstation to detect defects as soon as they appear is **source inspection,**[1] and is usually conducted according to a sampling plan that depends on the magnitude of the defect. This inspection permits immediate corrective action to strive for defect-free production. Through source inspection, the customer is outside of the quality management system.

Some firms will try to **inspect in quality** by letting defects occur and then finding them at a later point in the process with inspectors. This method complicates finding the root cause of problems because of the time delay between the occurrence of the problem and its detection. This approach allows a higher chance of creating defects that escape the process and are later found by the customer.

---

[1]See Reference 12 in the bibliography.

CASE STUDY

Inspection

An operator is performing a visual inspection on wafers that have a patterned resist coating from the photolithography process, and are ready to enter the etch process. The photo process uses a tool referred to as a track tool to apply the resist on the wafer, and then develops the resist after the photo step. This particular track tool is approximately 10 years old.

There are three main problems regularly found during inspection, which are listed by frequency of occurrence:

1. Stain on the resist coating of the wafer
2. Nonuniform resist coating
3. Contamination around the wafer edge

The inspection sampling plan depends on the circuit pattern that is put on the wafer, because data have shown that some patterns have more problems than others. For a difficult pattern, there is 100 percent inspection of all wafers, whereas for noncritical patterns, five wafers are randomly selected from a cassette of twenty-five wafers.

Problem analysis found the causes to be

1. Stains: improper suck-back from the nozzle during the spray of the developer, which causes drips on the resist and leads to stains. The problem occurs about once per week. When found, the operator adjusts the nozzle to correct the problem. Stripping the resist and reprocessing the wafer reworks the parts with the problem.
2. Nonuniformity: improper suck-back from the nozzle during the application of the resist. It also occurs about once per week, and requires a tool adjustment by the operator. The parts with the defect are reworked.
3. Contamination: a nozzle sprays a chemical to smooth the resist on the edge of a rotating wafer, and sometimes the nozzle is out of adjustment. This occurs about once every 2 weeks, and requires the tool to be adjusted by the operator. Defective parts are reworked.

Points to consider

1. Is this inspection a source inspection? Would it be better to inspect later in the process?
2. Is it acceptable to let the problems reoccur, since they only happen every week or two?
3. Are the problems acceptable as the tool is fairly old and possibly worn out?

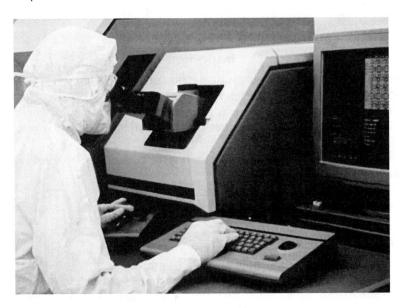

Inspection system for wafer defects.
Photo courtesy of Sematech Archives.

Inspection is a contentious subject in many firms, simply because reducing inspection appears to lower cost and increase the number of parts produced (fewer operators and inspection equipment required). With this approach, it is the customer who finds quality problems, which could be catastrophic for a firm.

To put the concept of inspection in perspective, the graph in Figure 5.3 describes how a process is improved as the defects in the process are found and reduced by corrective action. The high level of defects in Phase I is reduced through source inspection and immediate corrective action to correct the problem. This effort is value. As defects are reduced, the process moves temporarily through Phase II, which then permits a reduction in inspection because the defects have been corrected. Ultimately the process arrives at minimal waste in Phase III. The problems have been corrected and the source inspection is reduced to an audit to monitor the stable process.

Yield

Yield is the lifeblood of high-volume manufacturing and is the most often used indicator of product quality. Production yield, in percent, is defined as:

$$\text{Yield (\%)} = \frac{\text{Number of Good Product}}{\text{Total Number of Product Started}} \times 100$$

If a process has high yield, it means the process can produce parts that conform to the requirements and implies the product has acceptable quality. Note that yield

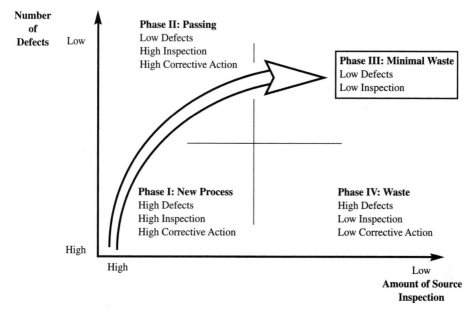

Figure 5.3 Minimal Waste with Source Inspection and Corrective Action

is an *indicator* of product quality, but it is not the actual product performance for the customer (known as **field performance**). A process can have high yield with poor field performance, which results from poor quality practices in the process.

 EXAMPLE 5.1 CALCULATING YIELD

If a workstation produces 100 wafers, and 2 are defective, what is the yield?

$$\text{Yield} = \left(\frac{98}{100}\right) \times 100 = 98\%$$

The yield is 98 percent, which means that based on this data, 2 wafers are expected to be defective for every 100 wafers produced.

How yield is calculated depends on where it is measured in the process. For instance, in a wafer fab, yield is measured by the station yield, which is the number of good wafers exiting a workstation, and the wafer fabrication yield, which is the wafer level yield at the end of the wafer process. These yields indicate how many wafers are

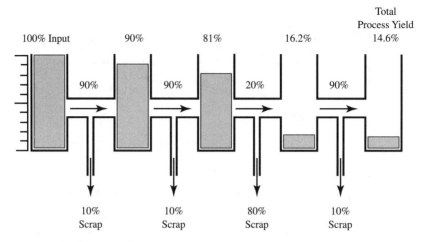

100% Input   90%   81%   16.2%   Total
Process Yield
14.6%

90%   90%   20%   90%

10%   10%   80%   10%
Scrap   Scrap   Scrap   Scrap

Figure 5.4 Total Process Yield with One Low-Yield Operation

scrapped in the process (e.g., broken or processed incorrectly). There is also wafer sort yield, which shows the percentage of die (individual chips on a wafer) that function properly after test. Yield should always be clarified to show where and how it is calculated, otherwise it could be misleading.

Yield that applies to the total process is calculated in various ways. The most accurate way is to multiply the yield of each station to obtain the total process yield. This is also known as line yield or accumulative yield (cum yield).[2]

For instance, a process with four operations that each has a yield of 90 percent has a total process yield of:

$$\text{Total Process Yield (\%)} = [(0.90)(0.90)(0.90)(0.90)] \times 100 = 65.6\%$$

An operation with poor yield reduces the total process yield. Figure 5.4 shows the effect of one low operation yield. The total process yield drops significantly, primarily because of one low-yielding operation of 20 percent. Note that the total process yield can never be higher than any individual station yield at an operation. This yield is a true representation of the total process, and is sensitive to poor performance from a single operation.

Another significant form of yield loss is **scrap parts.** Scrap parts are not reworkable and therefore thrown away, usually with a significant cost to the firm. If you work in manufacturing, sometimes the amount of scrap generated is unbelievable, often for such reasons as misprocessing (e.g., loading wrong software program),

---

[2]See Reference 17 in the bibliography.

CASE STUDY
Creating Scrap

A relatively new engineer in an older wafer fab performs an evaluation on a furnace used for oxide growth on a silicon wafer. The engineer runs some experiments with different furnace settings for gas flow into the furnace, and decides to change to a new setting for production. The operator is instructed how to set up the equipment with the new parameters, and told to run production.

After about a shift of production, a problem was found with the new furnace parameters. Because of the problem, twenty-four lots of wafers are scrapped. There are twenty-five wafers in a lot, and each wafer is valued at approximately $1000 (this is a true story, and it is not unique to any company).

Points to consider

1. Was it acceptable for the engineer to do an evaluation of the furnace settings?
2. How would you recommend doing the evaluation to increase the chance of finding correct furnace settings?
3. Was the process change implemented correctly? What standardized procedure was not followed correctly?

and other times due to the challenge of making a complex product. The biggest danger is when scrap becomes acceptable because it occurs so frequently.

Scrap loss in production is a strong indicator of equipment and process problems. A process must be immediately stopped for corrective action if a scrap part is made, or if there is the potential of making scrap.[3] The goal of the **line stop** is to determine the source of the problem, apply the team technical resources to correct it, and return the line back to production as quickly as possible with controls so the problem does not reoccur. Line stoppages for scrap loss always have priority for resources to resolve the problem.

Line stops can create a dilemma between a firm's need to have high-quality product, yet deliver product on time. If a customer shipment is due, some companies ignore quality to justify meeting the product ship date. This is not acceptable manufacturing quality.

In some cases, defective parts undergo **rework** and become acceptable. Rework avoids scrapping a part, but it is a costly effort. For wafer fab processes, few production mistakes are reworkable because there are not many ways to replace the materials that are introduced onto a wafer.

---

[3]See Reference 12 in the bibliography.

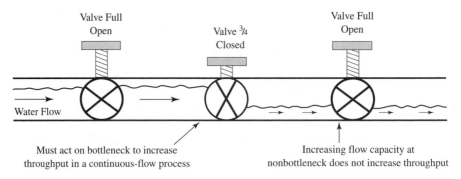

Figure 5.5 Bottleneck in a Continuous-Flow Process

Throughput

Throughput represents the actual quantity of parts processed per unit of time (for instance, each day). The equation for throughput is:

$$\text{Throughput} = \frac{\text{Quantity of Product Produced}}{\text{Total Operating Time}}$$

Throughput differs from capacity, because equipment capacity is what the equipment is capable of processing. A piece of equipment can have a 100-wafer-per-hour capacity with zero throughput (such as a tool stopped for lunch).

Throughput is defined for a specific operation or a total process. Most firms carefully monitor throughput at workstations to maximize their individual output. However, the firms then make the error of measuring throughput for the total process by how many parts are shipped by the manufacturing line in a given time (such as parts shipped per day). This confuses improvement, because it appears that just inputting more parts will lead to more output. This is not true.

Throughput for the total process is controlled by the slowest operation in the process, known as the **bottleneck** or **constraint.** In other words, once you know which operation is the bottleneck, then you understand what operation is **gating throughput.** The only way to increase the total process throughput is to improve the bottleneck throughput,[4] shown in Figure 5.5 for water flow in a pipe, which is the same concept as product flow in a high-volume manufacturing process.

Increasing throughput at nonbottleneck operations does not improve the total process output. However, note that as a process undergoes improvement, the location of the bottleneck is continually changing. Therefore no operation is immune from the need to improve, and teams must be cognizant of the total line improvement activity and where efforts are currently focused.

---

[4]See Reference 6 in the bibliography.

Throughput is important because it defines the quantity of product the firm can ship to customers. If customer demand is higher than throughput, then improvement of a bottleneck leads to increased throughput, which is good for the firm.

Cycle Time

Cycle time is the time allotted to make one piece or unit in production. Because parts flow through a high-volume process, total cycle time is the time interval it takes for one part to be produced at the end of the process. Cycle time is inversely related to throughput: decreased cycle time leads to increased throughput, shown in the following equation:

$$\text{Cycle Time} = \frac{\text{Total Operating Time}}{\text{Quantity of Product Produced}}$$

$$= \frac{1}{\text{Throughput}}$$

The process step or steps used for calculating cycle time can be an individual operation or any number of operations up to the total process. Example calculations are shown in Figure 5.6. Operation 2 has the longer cycle time of 2.5 minutes, versus 2.4 minutes for Operation 1. Interpret this as taking 0.1 minutes of processing time per wafer longer at Operation 2. Also note that the unit of cycle time is time per part, although it is commonly stated as just time.

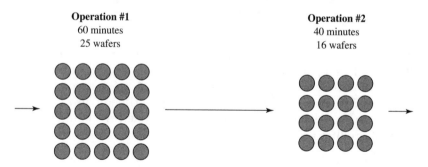

Operation #1
60 minutes
25 wafers

Operation #2
40 minutes
16 wafers

Figure 5.6 Calculating Cycle Time

$$\text{Cycle Time 1} = \frac{60 \text{ minutes}}{25 \text{ wafers}} = 2.4 \text{ minutes/wafer}$$

$$\text{Cycle Time 2} = \frac{40 \text{ minutes}}{16 \text{ wafers}} = 2.5 \text{ minutes/wafer}$$

$$\text{Total Cycle Time} = \frac{\text{Total Time}}{\text{Total Wafers}} = \frac{60 \text{ minutes} + 40 \text{ minutes}}{41 \text{ wafers}} = 2.44 \text{ minutes/wafer}$$

The importance of cycle time is its emphasis on the tangible quantity of time. If we reduce the time it takes to process parts (which is the cycle time), then the quantity of product produced will go up (this is throughput). This is true for the process where cycle time is measured. The goal is to decrease cycle time in order to reduce the time needed to deliver the product plus increase throughput.

How operators and technicians do their work has a direct effect on cycle time. Slow processing of parts through a workstation will increase cycle time and decrease throughput. This process could adversely affect a company's revenue (depending whether the poor cycle time is at a bottleneck and if the company's product is in demand). Cycle time was not commonly used in U.S. industry until the 1980s, and is still neglected in many firms in favor of yield and throughput at operations.

 EXAMPLE 5.2   CYCLE TIME AND THROUGHPUT

If 800 parts are produced at a workstation over an 8-hour shift, what is the workstation cycle time? If improvement then leads to a 20 percent cycle time reduction, what is the new workstation throughput?

Solution

The cycle time is calculated based on a throughput of 800 parts over 8 hours:

$$\text{Cycle Time} = \frac{8 \text{ hours}}{800 \text{ parts}} = \frac{0.01 \text{ hours}}{1 \text{ part}} \times \frac{3600 \text{ seconds}}{1 \text{ hour}} = 36 \text{ seconds/part}$$

The cycle time for this workstation is 36 seconds per part, based on data from these 800 parts. Thus a part requires 36 seconds to "cycle" through the workstation.

If cycle time is reduced 20 percent, then there is a 7.2-second reduction (36 seconds × 0.2 = 7.2 seconds). The workstation cycle time is now 28.8 seconds. To calculate the new line throughput, transpose the cycle time equation to solve for the quantity of product produced.

$$\begin{array}{l}\text{Quantity}\\\text{of Product}\\\text{Produced}\end{array} = \dfrac{\begin{array}{c}\text{Total Operating}\\\text{Time}\end{array}}{\text{Cycle Time}} = \frac{8 \text{ hours}}{28.8 \text{ sec/part}} \times \frac{3600 \text{ seconds}}{1 \text{ hour}} = 1000 \text{ parts}$$

$$\text{Throughput} = \frac{1000 \text{ parts}}{8 \text{ hours}} = 125 \text{ parts/hour}$$

This reduction in cycle time has led to an increase in throughput.

Tool setup time reduction for different parts is important to reduce cycle time.
Photo courtesy of Sematech Archives.

Reducing cycle time shortens the build time, which potentially increases throughput if it occurs at a bottleneck and results in reduced delivery time and lower manufacturing costs. For a wafer fab, a typical cycle time can be 4 to 6 weeks, which signifies the substantial opportunity in cycle time reduction.

5.1.1 Cycle Time Reduction: Series versus Parallel

An important way to reduce cycle time is to break work activities into series and parallel work. **Series work activity** occurs when actions are serial, one after the other. **Parallel work activity** occurs when actions take place simultaneously. If an operator can replace series activity with parallel activity at the workstation, then cycle time drops and throughput increases. There are usually many opportunities to make this type of improvement in daily work tasks.

 EXAMPLE 5.3   SERIES VERSUS PARALLEL WORK

For a tool setup, two actions are necessary: load a fixture that takes 30 seconds and load a software program that requires 90 seconds. If these must be done in series, what is the total setup time? If you improve cycle time by loading the fixture at the

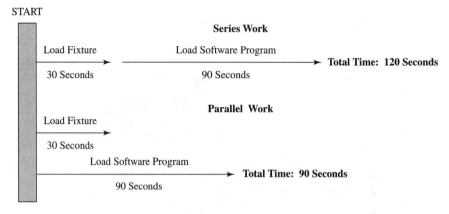

Figure 5.7 Series versus Parallel Work

same time the program is loading, what is the setup time? What is the percentage reduction in setup time by changing to parallel processing?

Solution

For series processing, the cycle time is the sum of the two steps, or 120 seconds. If the two steps are done in parallel, then the cycle time is equivalent to the longest time, or 90 seconds, as shown in Figure 5.7. The percent reduction in cycle time is:

$$\% \text{ Reduction} = \frac{\text{Original Value} - \text{New Value}}{\text{Original Value}} = \frac{120 - 90}{120} \times 100 = 0.25 \times 100 = 25\%$$

The result is a 25 percent reduction in tool setup time, which translates into a 25 percent reduction in cycle time at this operation. If this operation is a bottleneck, then process throughput increases by the same amount.

5.1.2 Cycle Time Reduction: Single-Piece Processing

Another way to reduce cycle time is to convert operations from batch processing to a **single-piece processing,** which is accomplished by reducing the lot size. Because parts are processed individually, there is less waiting time at each operation. The line becomes more balanced, WIP is reduced, and the cycle time is decreased.

Most conversions to single-piece processing are accomplished through installation of automated equipment that integrates stand-alone tools. The integrated equipment processes individual wafers in parallel steps for maximum cycle time efficiency. An example is a **cluster tool** that accepts multiple wafer cassettes, and

CASE STUDY

Parallel Steps at a Bottleneck

A throughput bottleneck exists at a particular thin film workstation in a wafer fab process, as evidenced by a high inventory of WIP waiting to be processed. There are two tools operated at this operation, and 70 to 80 percent of all lots in the wafer fab pass through this operation. Depending on the number of thin film layers on a wafer, lots process through this workstation two or three times.

Each time another product is run on the tool, it must be requalified by setting up new operating parameters. Operators work as a team to do as many parallel steps as possible to reduce the requalification time:

- Test wafers used in the requalification are gathered and characterized while the product is still running.
- These wafers are then loaded into the tool to be ready for processing while the last production lot is finishing its run.
- Once the requalification starts, then preliminary data are entered into the data entry screens and stored as partial data for verification later during the run.
- Measurement equipment is set up and calibrated while the run is in progress.
- Once the requalification run is complete, an operator takes the test wafers to a measurement station for data collection, while another operator collects measurements on the particle count in the tool.
- If all data pass, then the tool is ready to run a new lot.

By doing these steps in parallel, a requalification can be done in 20 minutes. If these steps were done in series, the requalification would take approximately 1 hour.

Points to consider

1. Does it make any difference whether this operation is a bottleneck or not?
2. Are these parallel activities complicated to implement?
3. When considering cycle time options, is it a factor that one of these tools costs around $6 million?

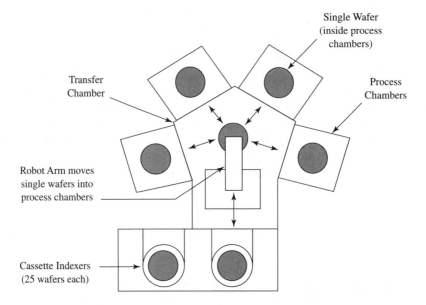

Figure 5.8 Cluster Tool Layout for Processing Semiconductor Wafers
Redrawn from *Solid State Technology*, January 1997.

moves wafers through various process chambers for simultaneous processing of individual wafers. A typical cluster tool is shown in Figure 5.8.

Process Flow

We visualize WIP in high-volume manufacturing as "flowing" through the process. The process flow for a particular product identifies the unique manufacturing steps for how WIP moves through the process during production. The actual equipment used for a product is defined in a floor control system that specifies what steps a product requires, and this information is communicated through software control or written documents.

A general outline of the major process areas in a wafer fab was described in Chapter 1. A complicating factor for the wafer-fabrication process flow is that wafers repeatedly cycle through process areas to add more surface layers. If manufacturing lacks strict adherence to standardized procedures, this creates more opportunities for errors that lead to scrap.

It may appear that operators and technicians have little control over manufacturing process flow. Although this is often true, understanding product flow gives more insight into a product and its particular needs for improvement.

The two basic forms of process flow are:

- Push process flow
- Pull process flow

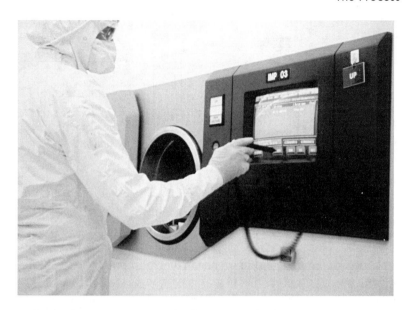

An operator checks the parts status in the floor control software system.
Photo courtesy of Sematech Archives.

The push process flow was the dominant flow method in many factories until the 1980s. A **push process** estimates the number of parts required from the process, and then inputs the quantity of components estimated to deliver these parts at the end of the line (with some upward factoring to account for yield loss). This process is inefficient because process bottlenecks and yield problems make it difficult to predict what is actually produced at the end of the process.

The major problem with a push process is it creates the false notion that simply putting more parts into a process will increase throughput (pushing parts through). However, we have learned that *total throughput is only increased by reducing the cycle time at the process bottleneck.* Putting more parts into a process without improving the bottleneck only increases WIP, which leads to trying to push even more parts through. This process becomes a vicious circle and is used as an expensive storage center for parts waiting to get through the bottleneck.

To improve manufacturing efficiency, some firms have implemented a pull process flow. A **pull process** evaluates customer demand, and then "pulls" the parts through the process based on this customer demand. It typically starts at shipping and works upstream through the process. A process can never pull faster than the bottleneck, as it is the gate to throughput.

For a pull process to function properly, operations are linked through use of **kanbans.** A kanban keeps operations from overbuilding by serving as a WIP communication link between downstream and upstream processes. With the use of kanbans, WIP buildup at an operation stops the line when the kanban is full and signals a

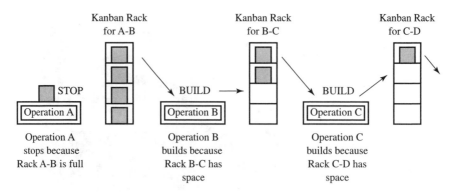

Figure 5.9 Simple Kanbans to Control Process Flow

process flow problem. There should be a sense of urgency any time a line stop occurs, with an appropriate team response (e.g., assist the operation with a problem, perform source inspection to identify bad parts while the problem is corrected). A possible kanban system based on a storage rack is shown in Figure 5.9.

When kanbans are first introduced into an inefficient process, there are frequent line stoppages. Kanbans stop operations that build faster than the bottleneck, because throughput is defined by the bottleneck. Parts can never move through a process faster than what is capable of moving through the bottleneck. Manufacturing with kanbans highlights the bottleneck to put focus on problem correction, cycle time reduction, and reliable equipment.

Inefficient firms that do not correct process problems and have inflexible equipment are reluctant to run a manufacturing process with true kanbans and minimum WIP. This situation requires a quick response to problems. If no WIP is in the line and problems are not quickly corrected, the process shuts down because of no throughput.

These firms actually want the exact opposite: lots of WIP in the line. The strategy is to work around the problems and somehow get the WIP through to final shipping, which negates the need to improve. Some firms also implement kanbans and then resort to various illusions such as buffer kanbans, which let the kanban size grow so that WIP can increase to cover up problems. Thus, problems are not highlighted and corrective action does not occur.

Specifications

Specifications are the manufacturing documents that specify the product and process requirements. Depending on the type of document, they are written and released by engineering. Manufacturing teams should provide input, review, and concurrence for engineering documentation (the amount of team input varies by company). In some cases, teams may be responsible for writing daily maintenance and operating procedures.

Wide variability exists among firms for their document release procedures. A central control group typically releases documents to manufacturing to ensure that the most recent version is on the floor. If a firm does not have good document control, it is possible for wrong documentation to cause incorrect work with costly rework or scrap.

Different types of specifications and procedures are used in manufacturing:

- Product specifications
- Process specifications
- Manufacturing procedures
- Maintenance procedures
- Calibration procedures and standards

Product specifications define the requirements necessary for the product to properly function according to all relevant customer requirements. They specify parameters, such as nominal dimensions and tolerances, material type, surface finish, purity, and so forth.

Process specifications specify the process parameters necessary for the product to be manufactured. The process specification is broadly interpreted in firms, because some prefer not to have engineering define the process (this is usually in older technology processes that do not have critical requirements). In this case, manufacturing will specify product parameters through the floor documentation.

Manufacturing procedures define the manufacturing requirements to produce the product. They are usually controlled by engineering and released to manufacturing through hard or soft copy documents known as recipes, run cards, and other terms. This floor documentation communicates the operating parameters at each workstation to the operators and technicians. In modern wafer manufacturing facilities, this information is downloaded from a software database into the equipment based on the product's part or lot number.

Maintenance procedures convey the requirements for maintaining the equipment in the factory. **Calibration procedures** are used to ensure that all test and measurement equipment are capable of making acceptable measurements. **Standards** are used to compare a measurement tool's output with known output to verify correct operation.

Training

Training is important because it is how knowledge is transmitted to manufacturing from the subject-matter experts: process developers, equipment manufacturers, manufacturing engineers, certified operators, and technicians. Manufacturing training encompasses classroom courses, hands-on training with the engineer, training by equipment suppliers, and training by experienced coworkers.

Training leads to flexible operators capable of adjusting to different manufacturing problems. Let's illustrate the benefits to flexibility from **operator cross-training** at different workstations.

EXAMPLE 5.4 OPERATOR CROSS-TRAINING

Part 1

There are two tools, Tool 1 and Tool 2, shown in Figure 5.10, and the operators are not cross-trained; there is no operator flexibility to work different tools. Tool 1 produces two parts/minute, and Tool 2 produces one part/minute. Where is the process bottleneck? What is the line throughput? How many parts from Operator 1 go into WIP storage after 1 hour?

Solution

The line throughput is set by the bottleneck at Tool 2, which is one part/minute. Fifty percent of the parts from Tool 1 move into WIP storage at the rate of one part/minute, while 50 percent are moved to Tool 2 for further processing. There are sixty parts in WIP storage after 1 hour, as shown in Figure 5.10.

Part 2

Now consider another scenario, shown in Figure 5.11, where operators are trained on both tools. Operator 1 can assist Operator 2 to increase line throughput, which may involve helping to load Tool 2, or prepping parts, or doing some activity in parallel to increase the throughput of Tool 2.

Let's assume that this team effort increases Tool 2's throughput to 1.25 parts per minute. In the worst case, let's assume that Operator 1's throughput also reduces by 25 percent to 1.5 parts per minute. What is the total line output per minute? How many parts are in WIP storage after 1 hour? What is the percentage increase in throughput and the percentage reduction in WIP over the previous condition?

Solution

The line output is 1.25 parts/minute, with 75 parts produced after 1 hour. There are 0.25 parts/minute entering WIP storage, or 15 parts after 1 hour. To calculate the percentage throughput increase and percentage WIP reduction, use the following equations:

$$\% \text{ Throughput Increase} = \frac{|\text{Original} - \text{New}|}{\text{Original}} \times 100 = \frac{|60 - 75|}{60} \times 100 = 25\%$$

$$\% \text{ WIP Reduction} = \frac{|\text{Original} - \text{New}|}{\text{Original}} \times 100 = \frac{|60 - 15|}{60} \times 100 = 75\%$$

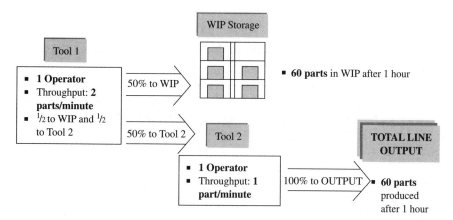

Figure 5.10 No Operator Flexibility (One Operator per Tool)

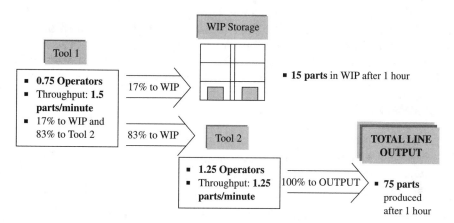

Figure 5.11 Operator Flexibility (Sharing Work Between Tools)

A 25 percent improvement in product throughput is attained with operator flexibility, with no capital cost to the firm, such as buying more equipment. There is a significant reduction in WIP of 75 percent.

Figure 5.11 demonstrates just one potential benefit to the firm from cross-training and operator flexibility. Multiskilled, flexible operators provide an added dimension for increasing manufacturing flexibility without extensive capital costs.

✳ CASE STUDY ✳
Advancing Operator Skills within a Team

An operator works in the etch process of a wafer fab. Etch is a complex process where selective areas of material are chemically removed from a wafer surface to form the circuit patterns and devices. There are six etch tools in this particular etch workbay run by this operator. Other tools are in the same bay operated by other team members, and they share work and assist each other depending on the workload and problems present.

At the beginning of each shift, the operator reviews the status of the six etch tools to determine which tools are available. Usually there is at least one tool down, and sometimes up to three or more. Because etch is complicated for submicron circuit dimensions, the most common reason for an etch tool to be down is a problem known as "etch rate." This is the speed at which the etch removal process occurs.

The operator has been trained by engineering to reestablish the proper tool parameters for etch rate problems. The operator loads a wafer monitor in the etch tool, and evaluates the etch parameters of temperature and pressure. After varying these parameters on the wafer monitor to attain the desired results, the operator resets the tool conditions and signs the tool back to production. If there is a problem, the operator talks with other operators or calls the on-site tool representative for assistance. Usually assistance is not necessary.

Points to consider

1. Is it reasonable for an operator to evaluate tool parameters and sign the tool back online?
2. What skills are necessary for this operator to correct the etch rate?
3. Who are the team members involved in solving this problem?

Process Control

Process control in manufacturing means that the process variables are controlled to produce predictable output and is accomplished using statistical process control (SPC) with team decision making to reduce unnatural process variation. This method statistically analyzes data to determine whether the process is in or out of statistical control. When a process is in control, it then behaves predictably, providing the confidence to produce good product.

SPC is a team tool used to assess process issues such as when a tool will become unstable, or if defective product could result from a process. The ability of its predictive power lies in statistics. It is no different than the ability of political pollsters to predict the outcome of an election before it occurs.

Another benefit from SPC is that a controlled process can be compared with the product specification to assess its ability to build parts that meet specification limits. This is known as **process capability**. A process is capable when it produces parts that repeatedly meet specification.

5.2 A HOLISTIC PROCESS VIEW

People typically focus efforts at their specific workstation, defining their process view within the confines of their work. This limited process definition is compounded by the complexity of modern manufacturing. Few people possess the necessary knowledge to comprehend the total process. This situation is dramatically different than a manufacturer of several centuries ago, when a skilled craft expert built the complete product (such as a chair maker who cut the wood, whittled the pieces, assembled the chair, and then sold it).

Specialized work tasks in a factory promote a narrow process view prevalent in manufacturing. This narrow view threatens productivity, because it creates excessive complacency about other operations to the detriment of the whole process. A person might reason, "I made my production numbers today, and the other operation is not my responsibility," or feel relieved that it is another area with quality problems.

Ultimately, manufacturing is an interdependent process that cannot be improved if its different parts are working against each other. A clear interdependency exists between all the manufacturing areas we have discussed, and they must be analyzed as individual parts contributing to the whole. If an operation has a yield problem, then cycle time increases, which reduces the throughput. If this operation becomes the process bottleneck, then the total line output decreases. As this operation improves, a new operation becomes the line bottleneck, requiring a new focus for improvement activity. Figure 5.12 highlights this manufacturing interdependency for different process conditions.

Each manufacturing operation is a subset of the total process, just as a finger is a part of the body. We must strive to reach a holistic process level, continuously viewing our individual work as just one small action within a total process. *A manufacturing **holistic process view** is essential for improving the entire process.*

Teams are the social and technical basis for a holistic process view. They provide a communication forum between people based on the technical needs at a workstation within the context of the total manufacturing line. They promote interaction with other teams across the entire process, developing technical diversity and competency. Teams emphasize humans as the special variable that is a part of the process yet able to effect change for process improvement.

(a)

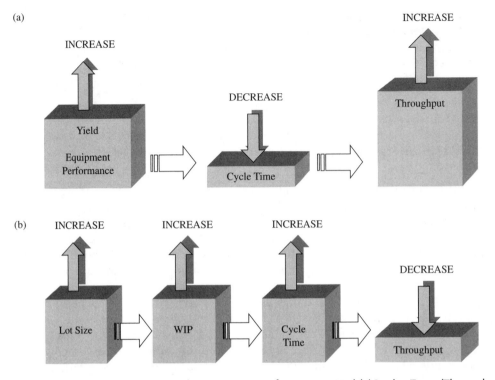

Figure 5.12 (a) Interdependency in Manufacturing: Yield/Cycle Time/Throughput; (b) Interdependency in Manufacturing: Lot Size/WIP/Cycle Time/Throughput

SUMMARY

High-volume manufacturing has extensive terminology to describe its technology. The workstation is where work occurs on WIP in the process, with a goal of minimizing WIP to expose problems for corrective action. Source inspection followed by corrective action is the most efficient method to ensure that product conforms to requirements. Yield is an indicator of product quality, whereas throughput represents the number of parts produced. Throughput is limited by the flow through the process bottleneck. Cycle time is the time allotted to produce one piece, and can be reduced through parallel and single-piece processing. Product flowing through a manufacturing line can be pushed or pulled through the process. A pull process in conjunction with kanbans is conducive for team-based improvement. The communication of product requirements occurs through specifications and procedures. Operator training gives added flexibility to respond to manufacturing needs. The process is controlled through the use of statistical process control (SPC). A holistic process view is promoted by teams and is necessary to optimize the total process.

IMPORTANT TERMS

Workstation	Series work activity
Equipment capacity	Parallel work activity
Setup time	Single-piece processing
Work in process (WIP)	Cluster tool
Floor control system	Process flow
Lot processing	Push process
Batch processing	Pull process
Inspection	Kanban
Source inspection	Specifications
Inspect in quality	Product specifications
Yield	Process specifications
Field performance	Manufacturing procedures
Scrap parts	Maintenance procedures
Line stop	Calibration procedures
Rework	Standards
Throughput	Training
Bottleneck	Operator cross-training
Constraint	Process control
Gating throughput	Process capability
Cycle time	Holistic process view

REVIEW QUESTIONS

1. What is the workstation in manufacturing?
2. How do you define *equipment capacity* and why is it important?
3. Describe WIP in manufacturing.
4. What does excessive WIP indicate in a process and why?
5. What is lot or batch processing in manufacturing?
6. Define *manufacturing source inspection* and *corrective action*.
7. Draw the plot that shows how source inspection and corrective action minimize waste in manufacturing.
8. Describe the different types of manufacturing yield.
9. What is the total process yield, and how is it calculated?
10. What are manufacturing scrap and rework?
11. How is a line stop used for quality?
12. What is throughput, and how is it affected by a constraint?

13. Define *cycle time* and show how it is calculated.

14. How does series versus parallel work effort affect cycle time?

15. What is the advantage for single-piece processing in manufacturing?

16. What are the two types of process flow? Discuss the pros and cons of each.

17. What is a kanban? Describe a simple kanban to control WIP.

18. List and discuss the different types of specifications and how they are used in manufacturing.

19. How does training affect manufacturing improvement?

20. What is process control in manufacturing?

21. Why is there opposition to a holistic process view in manufacturing?

EXERCISES

Inspection

1. HiTek Electronics produces a line of electronic equipment, including a small backup disk drive for computers. The small disk drive is portable and can be plugged into the printer port of a PC. In the final assembly of the disk drive, twenty-five employees work on the assembly line, using parts from a parts bin and small assembly tools. Each employee performs a different operation and then passes the assembly to the next person. The last two steps on the line are inspection and then shipping.

   Inspection includes an operational test of each disk drive and a visual inspection. If the drive fails inspection, it is placed at a bench where another employee reworks it. If the stack of drives at the repair bench grows too large for the operator to handle, then one of the regular operators is assigned overtime to reduce the repair backlog.

   A. What are the good points and bad points with this process?
   B. What type of product quality do you think is produced on this line? Explain your position.
   C. How would you improve this manufacturing process?

2. Refer back to Figure 1.4 in Chapter 1, which highlights the process flow for building a wooden cabinet with the quality inspection points. Given what you have learned about source inspection and corrective action, answer the following questions:

   A. The process for building the wooden cabinets is new and unstable. How should inspection be performed in this process? Where would you put the quality control (QC) points in this process to locate defects? What is critical for the process to improve? Show the inspection procedures on a flowchart.
   B. The process for building the wooden cabinets is stable and has undergone improvement. Cabinets produced from the process are high quality. Where would

you put the quality control (QC) points in this process to locate defects? Show this on a flowchart. Could the frequency of inspection be reduced?

Yield

3. The following data are collected for five operations:

Wafer Yield Data

	Number of Wafers Started	Number of Good Wafers
Operation 1	300	296
Operation 2	200	175
Operation 3	250	250
Operation 4	300	250
Operation 5	275	268

A. What is the yield at each operation? Which operation has the lowest yield and which operation has the highest yield?
B. Calculate the total process yield for all five operations.

4. At a wafer sort operation, five wafers with 100 die on each wafer are tested with the following number of good die per wafer:

Wafer Sort Test Data

Wafer #	Number of Good Die	Wafer Yield
1	88	
2	46	
3	83	
4	94	
5	89	

A. What is the yield of each wafer based on these data?

B. If you want to improve yield, which wafer would you evaluate first for defects?

Cycle Time

5. You work at an operation that has a process yield of typically 100 percent. On a good day (meaning all the equipment is up), your team can process twenty-four lots of wafers (twenty-five wafers in a lot) in a 12-hour shift.

A. What is the cycle time for parts processed through this operation?

B. There is a problem with the quality of incoming parts, and yield drops to 75 percent (the remaining 25 percent of wafers are scrapped after processing through your operation). What is the new cycle time? What is the throughput for one 12-hour shift?

C. The team has found a way to rework defective wafers, but each rework takes 15 minutes per lot in addition to the regular processing time. What is the new cycle time for twenty-four lots of wafers?

D. What is the best solution for correcting this problem?

6. You have collected the following data for three operations that work in series (Operation 3 follows 2, which follows 1):

Data by Operation

	Operation 1: Photo	Operation 2: Etch	Operation 3: Furnace
Time Period	One 12-hour shift	One 12-hour shift	One 12-hour shift
# of Wafers Produced	500	300	475

A. What is the cycle time for each operation?

B. What is the throughput of this process (Operations 1, 2, and 3 combined)? Which operation is the bottleneck?

C. How is it that Operation 3 is producing more parts than what is supplied to it?

Series versus Parallel

7. You are making breakfast in the morning and want to prepare oatmeal, muffins, and orange juice in the shortest time possible. Lay out a simple line diagram for each of the following conditions (similar to Figure 5.7, with 15-second intervals), showing the total time for each condition:

A. All activities in series.

B. Activities in parallel to the extent possible (e.g., if you are preparing oatmeal, then that occupies your time; however, you can do other activities in parallel while the oatmeal is in the microwave).

Oatmeal:

Prepare oatmeal	45 seconds
Microwave oatmeal (first time)	2 minutes
Stir oatmeal	15 seconds
Microwave oatmeal (second time)	1 minute
Put on sugar	15 seconds
Pour milk	15 seconds
Carry to table	15 seconds

English Muffin:

Slice muffin and put in toaster	15 seconds
Toast muffin (first time)	1 minute (then pops up)
Toast muffin (second time)	1 minute (then pops up)
Butter immediately	30 seconds
Carry to table	15 seconds

Fresh-Squeezed Orange Juice:

Get three oranges from refrigerator	15 seconds
Slice oranges	15 seconds
Squeeze in juice machine	2 minutes
Pour in glass and take to table	15 seconds

Holistic Process

8. Your team is analyzing a manufacturing line to assess problems. For each of the different scenarios, describe how this could cause a manufacturing problem. What would you recommend as a solution to correct each problem?

   A. The throughput is low for the process. You find excessive WIP stockpiled in front of an operation with a low yield.
   B. The floor control department has informed your team that to increase the line throughput, they propose increasing the lot size to move more parts through the process.
   C. A particular workstation has over three times the cycle time of the other operations in the process. You find that the tool availability is 50 percent.

6

✳ SOURCES OF PROCESS WASTE

We become accustomed to waste and accept it as a natural by-product of a process. Effective improvement activities eliminate process waste. To succeed in a competitive market, our goal is to reduce waste to increase value.

OBJECTIVES

After studying the material in this chapter, you should be able to:

1. Explain how to increase value in a manufacturing process.
2. List the seven sources of waste in manufacturing.
3. Explain the negative consequences to manufacturing from each type of waste, and give a manufacturing example.

6.1 VALUE VERSUS WASTE

Continual improvement dictates that we never accept a manufacturing process in its present form. It may seem contrary to achieving a holistic process view, but we must continually question the existence of each part of a process and its value to the product to ultimately improve.

Once we accept the need to question every aspect of a process, we must then ask ourselves, "Which efforts are adding value, and which are waste?" Waste that is eliminated piece by piece is the challenge for increasing value in a process. To a human, just the act of doing something implies it is important and therefore of value. This belief is confused with the concept of improvement, because it makes us think all work is important. To avoid this trap, study every level of detail in an operation to find the efforts that are not necessary and are therefore waste.

*The only way to **increase value** in a process is to continuously seek out and identify waste, and then eliminate it.* Each time waste is removed, value increases and the process becomes more efficient. This ongoing effort to identify and remove waste is the foundation for continuous improvement. Through continual improvements, the process becomes more competitive with lower cost, higher quality, and shorter delivery.

6.1.1 Identifying Waste

Some sources of waste are obvious and are quickly eliminated. Others are hidden in the process and are difficult to find. Furthermore, other forms of waste are hard to identify because of predetermined notions about how something should be done. We easily fall into the trap of thinking, "This is the way it has always been done" or "This is the way I was told."

To **identify waste** and thus eliminate it, we must continually reflect upon our work by asking the following questions:

- What is the minimum effort required for the product to work?
- Is this activity needed? Why?
- Are there other simpler or more straightforward methods?

Never limit efforts to only certain sources of waste. To eliminate manufacturing waste must become the guiding principle at work. There is no such thing as minor waste—waste must be eliminated regardless of how it is categorized.

Even harder than questioning about sources of waste is dealing with individuals who resent someone else questioning their work. Some people have mental blocks to improvement, opposing change for reasons such as lack of recognition or insecurity about their job. No workers are immune to this so-called mental block.

The most effective way to change and work with people with mental blocks is through positive team accomplishment. Achieving good results creates a group standard that provides a sense of accomplishment and security not easily available to individuals acting alone. The team should focus on improvement while allowing time for nonbelievers to see the results and decide to participate. *The key is evolution, not revolution.*

The effort to eliminate waste is fundamental to continual improvement. Eliminating waste helps to move the process toward the three goals of competitive manufacturing: lowest cost, highest quality, and shortest delivery. The precise amount of waste in any process can never be determined, and is actually not important to know. What is critical is the effort to identify and eliminate waste, thus improving manufacturing's ability to compete.

6.2   SEVEN SOURCES OF WASTE

There are seven major sources of **manufacturing waste.** They are interrelated and form the basis from which other forms of waste are derived. We now analyze the manufacturing process from the standpoint of waste, thus gaining a deeper understanding of the process interdependencies and the knowledge of how to continually improve.

The seven sources of manufacturing waste are:

- *Nonproductive work*
- *Idle time*
- *Floor layout*
- *Lot size*

- *Product defects*
- *Overproduction*
- *Off-line manufacturing control*

Nonproductive Work

Nonproductive work is waste because an effort made in manufacturing is not adding value to the product. This form of waste is difficult to identify because workers appear busy, sometimes even overworked, but are doing nonproductive work.

Typical examples of nonproductive work found in manufacturing are:

- Making a tool setup for a product that takes longer than what is possible
- Repeatedly inspecting parts to find defects and not taking corrective action
- Making charts and attending meetings when there is no effort to improve
- Working inattentively with a lack of discipline

Idle Time

Idle time occurs when a resource is not being used to add value to the product. Waste from idle time is relatively easy to detect in a process by doing a walk-through of the production areas to look for waste such as stockpiled parts or idle equipment.

The three primary ways for idle time waste to occur are:

- Idle humans
- Idle equipment
- Idle parts

Idle humans are considered as waste because work is not done. Idle time waste does not conversely mean that a human has to be working every instant. There are legitimate times when people are not working, such as during breaks or discussions concerning a process condition. Such activities differentiate humans from other process elements and contribute to manufacturing efficiency. Actions such as avoiding process improvement needs, gossiping with other employees, or wandering around the process with no work, however, are all forms of idle time and waste.

Idle equipment means the equipment is not operating at the maximum time possible. This downtime leads to decreased product throughput. On the other hand, neglecting equipment maintenance to run equipment continuously is also waste. Equipment availability will drop and create production problems.

Idle parts occur when throughput bottlenecks restrict parts from flowing through the process, which is evident by WIP buildup in front of the bottleneck operation. Bottlenecks limit the ability to efficiently make product, because some portion of the manufacturing effort is going toward building idle parts that are sitting on the production floor.

✳ CASE STUDY ✳
Wafer Sort

After a semiconductor wafer has completed all fabrication steps, it undergoes a test operation known as wafer sort to identify which die (chips) on the wafer are defective. The wafers are tested in an automated piece of equipment called a wafer tester. An operator places a cassette of twenty-five wafers in the tester, loads the proper software program for the particular wafer, and conducts wafer alignment. The actual test takes approximately 3 hours for twenty-five wafers, and one operator runs four testers.

Wafer alignment in this tester involves operator judgment. Once the tool-handling mechanism moves a wafer from the cassette to the test fixture, the operator positions the wafer so that an array of very fine diameter pins located on a probe card can make contact with pads on each wafer die. The operator uses a joystick and video screen to make fine adjustments to the wafer position. The requirement is for all probe card pins to make good contact with each contact pad. Approximately once per day, probe card pins will not align properly on any tester.

If after several attempts the operator cannot align the pins to the contact pads, then the operator removes the probe card and replaces it with another card that has had the pins straightened. Probe card pins are fragile and easily bent. No new probe cards are available. Probe cards with bent pins are sent to an on-site repair center that uses a fixture to straighten the pins. In most cases, however, the repaired probe cards that are returned to the tester still have pins that do not align properly. After the operator tries to install a repaired probe card and it still does not align, then the operator shuts down the machine and calls engineering.

Points to consider

1. Does it sound normal for probe card pins to bend so often (once per day per tool)?
2. Why are the repaired probe card pins still not properly aligned?
3. What is engineering doing to bring the tool back online and error-proof the problem?
4. Could a team approach be more effective for solving this problem?

Floor Layout

Manufacturing floor layout can create waste. The two primary floor layouts for manufacturing processes are:[1]

- Process-oriented layout
- Product-oriented layout

A **process-oriented layout** groups similar tools together in the factory. This layout was common in the United States from the 1950s to the early 1980s, and was used because of the perceived ease in maintenance, facility support, training, and supervision when similar tools were in close proximity to one another. It also fit well with large lot sizes that were prevalent at the time, as a group of tools could simultaneously work on a large lot of parts (creating the impression of efficiency).

Examples of waste from a process-oriented floor layout include:

- Transportation and WIP waste because operations are isolated and farther apart
- Increased cycle time because parts wait longer
- Risk of reworking or scrapping a large quantity of parts stored in WIP if a customer requests an in-process change (known as an *engineering change* or *EC*)
- Lack of interaction between operators and technicians at different operations, making it less conducive for teams

An example of product flow through a process-oriented floor layout is shown in Figure 6.1. This type of floor layout often leads to confusing product flow, and

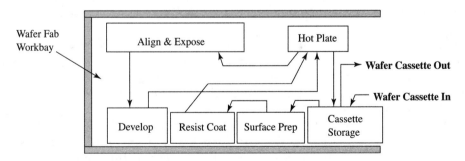

Figure 6.1 Process-Oriented Layout for Stand-Alone Wafer Photolithography (1970s era)
Note: Every arrow requires human transport to load and unload a cassette of wafers into and out of the tool. Each cassette carries up to twenty-five wafers.

---

[1] See Reference 15 in the bibliography.

A service-chase bay between work-bays provides access to tools, facility utilities, and supply piping for process chemicals.

illustrates a good first test for assessing the efficiency of a manufacturing process, described as a visual flow test. If you have difficulty following the product flow in a process, with parts moving in an illogical flow, then the process probably has poor floor layout and is wasteful.

Product-oriented layout (also known as group technology) in manufacturing emphasizes the needs of the product. It groups together the tools required to build a product by creating integrated lines, usually by combining several tools and process steps. Equipment is coupled with automated conveyors and position sensors, and then linked with software control.

Wafer fab floor layout is typically organized around workbays due to technical needs, such as contamination control and the delivery of chemical and gas systems. Access for maintenance personnel to repair and service the equipment is from service-chase bays located between the workbays.

Advances have been made toward product-oriented layout in wafer fabs through integration of multiple operations, such as that achieved with a tool called a *track system*. A product-oriented manufacturing layout for the same photolithography process previously shown in Figure 6.1 is now shown in Figure 6.2a and 6.2b using examples of two different integrated track systems.

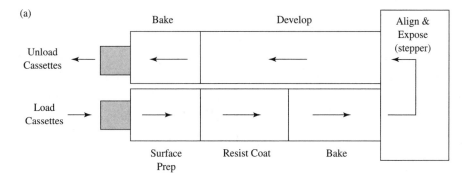

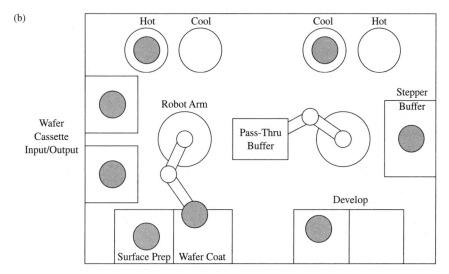

Figure 6.2 (a) Product-Oriented Flow for Photolithography with a Track System; (b) Product-Oriented Flow for Photolithography with a Robotic Track System

Note: The operator loads and unloads up to four cassettes of wafers. All process steps occur with single-wafer, parallel processing. Wafers move between operations on conveyors with wafer handling systems. Operations are software integrated and controlled, permitting one operator to run multiple systems.

Process steps before and after the align and expose tool (or stepper) in Figure 6.2a and 6.2b are integrated into the track system. The track tool is then attached to the stepper, and wafers process through the integrated operations with single-piece flow and parallel processing. *System integration of discrete process steps leads to substantial reductions in cycle time and increased product throughput.* Human intervention to load wafer cassettes occurs only at the load and unload station of the track tool. Note how the flow of the process makes logical sense with respect to the production needs of the product. As mentioned, a visual flow test indicates process efficiency.

EXAMPLE 6.1   EFFICIENCY IN PHOTO

The process-oriented layout for wafer photolithography shown in Figure 6.1 can process twenty-five wafers (one lot) in 2 hours, whereas the product-oriented track system can process one lot of twenty-five wafers in 45 minutes. What is the manufacturing efficiency of the process-oriented flow of Figure 6.1?

Solution

Assume the product-oriented track system is the theoretical minimum effort.

$$\% \text{ Efficiency} = \frac{\text{Theoretical Minimum Effort}}{\text{Actual Effort}} \times 100 = \frac{45 \text{ minutes}}{120 \text{ minutes}} \times 100 = 37.5\%$$

The efficiency of the process-oriented flow is 37.5 percent when compared relative to the product-oriented flow. If this is a bottleneck, then increase throughput by improving the track system. At the same time, the inefficient process-oriented flow can help balance the line at peak periods if it remains available to production (e.g., by running older product to free up equipment for new products).

Product-oriented production reduces waste and provides an integrated equipment system for assisting the team to improve both the equipment and the process. It breaks down the human barriers created by divisions based on discrete tool sets—barriers that are contrary to team-based improvement activities.

Processing multiple part numbers through a product-oriented line creates the risk of inefficient equipment usage. This problem occurs when equipment is inflexible and requires long setup times for each product. To address this concern, quick product changeover for tool flexibility is important for product-oriented floor layout. Reducing setup times requires a mental shift in viewing work activities. Question each step in the setup, ask why it is done, and how it could be done faster and more efficiently.

Lot Size

As we have seen, one lot is a group of parts processed together. The lot size is determined based on different needs. The five main considerations for determining lot size are:

1. Decreased lot size reduces WIP with a corresponding increase in throughput.
2. Decreased lot size highlights production inefficiency problems exposed by the reduction in WIP.

CASE STUDY
Reducing Setup Time

Modern efforts to reduce setup time can be traced to Mr. Shigeo Shingo, a Japanese industrial engineer involved in manufacturing improvement in many Japanese firms.[2] As an example of a setup time reduction project, Mr. Shingo found a way to reduce an 8-hour setup time to exchange die and tools for a piece of equipment down to only 58 seconds. He termed his approach "single minute exchange of die" or SMED, in which he claimed that any equipment setup from one condition to another could be done in less than 10 minutes.

The eight major steps to setup reduction using SMED are

1. Separate all process steps into internal and external steps. Internal steps require the machine to be stopped during setup, whereas external can be done in parallel while the equipment is running.
2. Convert process steps from internal to external.
3. Standardize functional steps for setup work.
4. Replace time-consuming steps (e.g., screwing bolts) with time-saving features (e.g., clamps).
5. Use standardized fixtures.
6. Implement parallel operations.
7. Eliminate adjustments.
8. Use equipment automation.

Points to consider

1. How many of these steps were used in the Case Study: Parallel Processing at a Bottleneck found in Chapter 5?
2. What is the benefit of converting process steps from internal to external?

3. Increased lot size reduces product variability from run to run at a tool (since more product is exposed to the same tool conditions).
4. Increased lot size achieves economy of scale during processing (e.g., loading five at a time is better than loading one at a time).
5. Increased lot size attains transport efficiency if operations are far apart.

---

[2] See Reference 12 in the bibliography.

A standardized fixture procedure reduces tool setup time.
Photo courtesy of Ion Implant Services, Inc.

Large lot sizes increase WIP in the process and contribute to waste by covering up manufacturing problems. Reducing waste by decreasing lot size is a balance between the five considerations listed here for determining lot size. In most cases, product lot size is not a variable over which operators or technicians have control. There may be legitimate business or technical reasons for not decreasing a lot size to the smallest possible size; however, when the choice is not detrimental to other process conditions, manufacturing should strive to reduce the lot size.

Lot size affects the flow of parts through the process. Large lots move through operations like a train that keeps starting and stopping, disrupting the flow of the cars. When a large lot is at an operation, there is excessive work to do. When the lot moves on to the next operation, there is no work left. This factor slows the flow of parts through the process, which then leads to increased cycle time because most parts wait at operations instead of being processed, as demonstrated in Figure 6.3.

Observing Figure 6.3, notice how the parts in the single-piece flow process are spread out through the four process steps of A, B, C, and D. When a part is done at a process step, it moves to the downstream step without having to wait for other parts in a lot. As shown in this example, single-piece flow of parts reduces cycle time from 16 minutes to 7 minutes (56 percent improvement), and increases throughput.

Large lot sizes create excessive WIP that hides process problems, thus leading to more waste in a process. For example, if an operation builds up a large quantity of

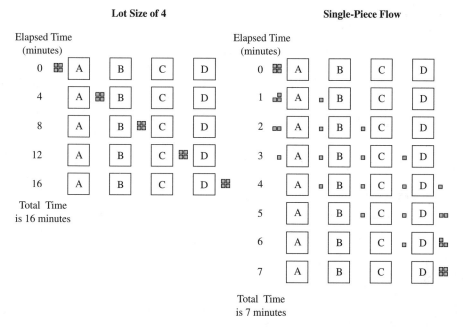

Lot Size of 4   Single-Piece Flow

Elapsed Time   Elapsed Time
(minutes)   (minutes)

Conclusion: Single-piece flow reduces total cycle time from 16 minutes to 7 minutes

Figure 6.3 Lot versus Single-Piece Flow[3]

parts, and then the tool breaks down, downstream processes are able to keep work-ing because of the large WIP buildup. Some firms actually desire this situation, be-cause it appears to keep the line producing parts even though a piece of equipment is down; however, it is the wrong approach. Problems such as equipment downtime cannot be ignored. Small lot sizes with reduced WIP force manufacturers to address the root problems in production.

Small lot sizes not only balance the flow of parts in a process, but also facilitate "communication" between operations. There is no ability to hide problems behind excessive WIP. Teams have no choice but to work together to improve the process. Lagging operations that do not improve become new bottlenecks, which is evident to everyone because of the WIP buildup at the operation. This situation forces im-provements.

In some manufacturing processes, the lot size is one. For instance, a bottling firm has a continuous flow of bottles processed through the factory, and each bottle is cleaned, labeled, filled, and sealed independently.

---

[3] See Reference 15 in the bibliography.

Product Defects

The creation of defects is a large source of waste. Product defects occur for many reasons: human error, improper equipment settings, equipment failure, problems with incoming parts, and so forth. Many defects occur because of lack of discipline for following the standardized work procedures. Another large source of product defects is inadequate process repeatability. The scourge of manufacturing defects is that they must be found, analyzed, repaired, or scrapped. If not, they will be passed on to customers. This defect management must be done while the firm is trying to manufacture and ship acceptable product.

Some examples of waste associated with defects are:

- Inspection with no corrective action
- The time to separate and keep defective parts from good parts
- The time necessary to analyze defects
- The cost to rework a defective part and make it good
- The cost of disposing scrap parts

Inspection with no corrective action is waste because its only purpose is to find defects, with no corrective action taken to correct the problems. Problems continue indefinitely, along with the associated inspection activities. A high likelihood exists of defects escaping to customers, because it is impossible to find 100 percent of all defects that are randomly generated in a process.

Conversely, if manufacturing personnel search for defects and also correct the problem at the operation, then this is value-add. We have previously defined this as source inspection. With source inspection and corrective action, defects will reduce and quality improves.

Overproduction

Overproduction occurs when product is built that is not required by customer demand—either the external customer who buys the product or the internal customer at each process operation. Overproduction stems from operations producing at capacity regardless of the downstream demand. It is visually evident when large amounts of WIP are stored between operations.

Overproduction is waste, because effort is expended to produce product that will only be stored on the manufacturing floor. This work is unneeded and disrupts the product flow through the process. Examples of overproduction are:

- Additional handling of materials
- Operating equipment to build parts that are only stored
- Inventory costs to buy parts and floor space to store the WIP
- Confusion and lack of purpose at operations

A balanced line evens out production with respect to cycle time, which balances the flow through the process while minimizing WIP and in turn reduces overproduction. To balance a line so that parts flow through the process evenly is also known as *process leveling,*[4] or leveling the product flow through each operation. When a team commits to balancing the line, then all factors affecting throughput are considered, including product defects, equipment availability and capacity, setup times, resource utilization, lot size, and WIP buildup. As the team improves these areas, the line will balance.

Line balance improvements are made by focusing on bottlenecks while concurrently evaluating all areas of a process. Once a line is roughly balanced, then the team continues to reduce the cycle time by seeking new bottlenecks to optimize. The team recognizes that a manufacturing line is dynamic due to interaction between the elements. What is optimal today may very well become the bottleneck tomorrow.

Overproduction is an insidious form of waste that is difficult to identify as a problem, because people and equipment are working (and in some cases, working hard to produce more than is capable of being processed). In this sense, overproduction is related to waste from nonproductive work. Some different sources of overproduction are shown in Figure 6.4.

Off-Line Manufacturing Control

Off-line manufacturing control occurs when managers and technical people use meetings to disseminate process information and make off-line decisions concerning day-to-day production needs. Meetings require a large support staff (engineers, staff assistants, coordinators, technical interface people) to acquire and bring

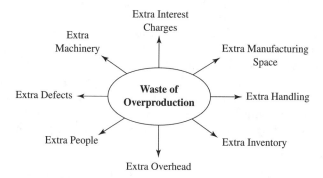

Figure 6.4 Waste from Overproduction
Redrawn from "The New Manufacturing Challenge," K. Suzaki, 1987.

---

[4] See Reference 12 in the bibliography.

process information into the meetings. Sample topics for meetings include: what product to run, yield management, how to correct equipment problems, how operators will perform their work at an operation, and line shutdown problems.

The inefficiency from off-line manufacturing control derives from people making critical process decisions who lack detailed knowledge about the process, which creates an insatiable need for information and promotes an extensive amount of nonproductive work by technical people. Examples of waste created by off-line manufacturing control are:

- Engineers become coordinators or planners who relay information between manufacturing and management.
- Operators and technicians input data into computer terminals or on data sheets, with no action to improve the process. Data are used for presentations in meetings.
- Management conducts lengthy status meetings to review information and provide direction. Much of the information is outdated, after the fact, and liberally interpreted to justify predetermined decisions.
- Improvement activities from the off-line meetings do not address the real manufacturing problems.

✳ CASE STUDY ✳
Meetings

An engineer is hired to assist the improvement activity in the thin film process area of a wafer fab. This engineer will work with the operators and technicians, and will report directly to the manufacturing supervisor. To communicate the status of all product in the area, the engineer calls a daily meeting at 7:00 A.M. during shift changeover. All operators and technicians in thin films are required to attend the meeting, which lasts approximately 20 minutes. It is held in a cleanroom so that people do not have to remove their protective clothing.

The engineer presents a chart showing which product has been processed, where the bottlenecks are located, and the important lot numbers for the next shift. The chart is available on the computer at the thin film workbay, and is regularly monitored by operators to identify the critical lots in the process.

Points to consider

1. Instead of having all the operators come to the engineer for a meeting, could it work for the engineer to go to the workbay and meet with the operators?
2. Is it reasonable to present a chart that is already available and monitored by the operators during their work?

Managing production through off-line control became prevalent in manufacturing firms during the 1960s and 1970s. Manufacturing was noncompetitive and encountered problems during this period. The rationale was that most manufacturing problems were unsolvable in manufacturing and needed management action. Given the depletion of skills and resources in manufacturing that were common at the time, it was true that manufacturing could not solve its own problems. However, the solution required solving the root problem in manufacturing, not moving control outside of manufacturing.

Ways to improve and control manufacturing processes have evolved since the late 1970s, and manufacturing personnel have been forced to address the importance of teams to achieve more process efficiency. This is not to say that all teams function properly, but effective teams focus skills and resources in manufacturing, at the source, giving organizational support for making critical day-to-day decisions to improve manufacturing.

SUMMARY

Waste must be identified and eliminated in a process, which leads to increased value in the product. There are seven major sources of manufacturing waste. Nonproductive work is effort that does not add value to the product. Idle time waste occurs through idle humans, equipment, and parts. Floor layout contributes to waste when it is designed around the process instead of the product. Flexible equipment with short setup times is important for product-oriented floor layout. Large lots increase WIP and hide manufacturing problems. The determination of lot size is a balance between different process factors, and is not typically controlled by the operator. Product defects generate significant waste. Overproduction is waste because effort is expended for product that sits on the floor as stored WIP. Off-line manufacturing control generates extensive work that detracts from the focus to resolve process problems.

IMPORTANT TERMS

Increase value
Identify waste
Manufacturing waste
Nonproductive work
Idle time
Idle humans
Idle equipment
Idle parts

Floor layout
Process-oriented layout
Product-oriented layout
Lot size
Product defects
Overproduction
Off-line manufacturing control

REVIEW QUESTIONS

1. How does someone increase manufacturing value?
2. Why is waste difficult to identify?
3. List the seven sources of manufacturing waste, and give an example of each.
4. How has the change from process flow to product flow benefitted equipment integration?
5. What are the five considerations for lot size?
6. How does a large lot size hide problems and how can a reduced lot size improve communication?
7. When is manufacturing inspection considered waste?
8. How does overproduction lead to waste, and what is done to control this?
9. Give an example of how off-line manufacturing control leads to waste.

EXERCISES

Waste

1. Identify the waste in the following situations. How could this waste be improved?
   - A. At the end of a run, the cassette of completed wafers is put on a cart. The cart can hold ten cassettes. The cart is moved to the next operation when ten cassettes are completed and loaded on the cart.
   - B. The preventive maintenance (PM) activity on the tool was scheduled for 9 months ago, but production needs were too critical to take the tool down for the required 24 hours. The tool is experiencing alignment problems, and has scrapped four lots of wafers in the past 2 days (one lot is twenty-five wafers).
   - C. A particular operation is a process bottleneck. The operation stops every shift for 1 hour while the operators are at lunch.
   - D. The supervisor has a 30-minute meeting with all operators at the beginning of the shift. The operators discuss the status of their work.
   - E. The process technician counts the status of all parts in the work area at the beginning of each shift and provides this information to the supervisor.
   - F. An operation builds all the parts possible every shift, even though the next downstream operation can only build half as many parts per shift.
   - G. Defective parts are made occasionally at a workstation, but there is no effort to find them because the operators are sure the test operation will find them.
   - H. Defective parts are made occasionally at a workstation, and the operators thoroughly inspect all parts to find them. When a defect is found, it is immediately reworked and moved on to the next process.

2. For the case study in this chapter titled "Reducing Setup Time," give a short explanation of how each step contributes to reducing cycle time at a workstation.

Lot Size and Product Flow

3. This exercise is a simulation to show how lot size can create waste from the flow of product through a process. It requires the participation of the class. Form into two teams: the LL team (large lot) and the SP team (single piece).

   Step 1: There are two separate processes, each operated by a team:

   LL team (large lot): Parts processed with a large lot size of four parts.

   SP team (single piece): Parts processed with a single-piece lot size of one part.

   Step 2: Each process consists of four operations with an "operator" at each step. The operators can sit next to each other at a table. Each step requires a 10-second wait to process the part, which is signified by operators raising their hand while "working." There is no transport time between operations.

   Step 3: Provide each process (LL and SP) with sixteen parts at the first operation (use any convenient object as a part, such as pieces of paper or cardboard).

   Process LL: The lot of four parts can only move as a group between operations (once all four are complete at an operation, which is 40 seconds).

   Process SP: A part moves to the next operation immediately after the 10-second work period.

   Step 4: A timekeeper tells the operators when to move parts after each 10-second block, and records the total time to process all sixteen parts through each process. This can be done on a board or flip chart.

   A. Compare the total amount of time it takes to process the parts in SP versus the amount in LL. Which process is faster?

   B. Calculate the percentage difference in total time. Why is there a difference in total time between these two processes?

4. Now change process SP in the previous problem so that one operation is down 50 percent of the time (every other 10 seconds one of the operations stops and no parts flow through, letting excess WIP build up). Rerun the simulation and compare the total processing time between LL and SP.

   A. Which process is faster? Why is this different from the first run?

   B. Can you use this information to explain why someone might think large lots are better if problems occur in the process?

   C. What observations can you make about the challenges of implementing process changes such as single-piece flow into a complex manufacturing line?

7

✳ IMPROVEMENT

Improvement is a way of life, a continuum over time. It is fundamental to existence as a means to continually increase efficiency for using the resources in our life system. Any process that does not improve over time is headed toward its eventual demise, unable to compete with other processes that improve.

OBJECTIVES

After studying this chapter, you should be able to:

1. Discuss how a new manufacturing process improves and how teams contribute to this improvement.
2. List the three levels of improvement, and explain how they affect a manufacturing firm's competitive position. Discuss which level is most effective for manufacturing improvement.
3. List and discuss the three classes of equipment.
4. Draw and explain the bathtub curve for equipment performance.
5. Explain chronic and sporadic equipment failure and how these aid in improving equipment performance.
6. List and discuss the six reasons for poor equipment performance.
7. Given the OEE equation, calculate and interpret the results for where to focus improvement effort.

7.1 MANUFACTURING IMPROVEMENT

Manufacturing improvement is the identification and analysis of manufacturing variables to understand their interaction, followed by optimization to achieve the three goals for competitive manufacturing: lowest cost, highest quality, and shortest delivery. Some variable interactions are well understood, but others are not clear. As an example of unclear interaction, intermittent defects occur on an automated tool and no one understands why. Given all the different possible manufacturing variables, their levels, and how they interact, you can appreciate the complexity of improving manufacturing.

Improvement occurs over time if sufficient effort is put into the process. A process rarely starts efficiently, whether it is a manufacturing process or a natural process. The trend is always for a process to begin inefficiently, reflecting inexperience and lack of knowledge about the process itself. When effort is put into the process, then it improves through the learning curve. No manufacturing process improves by itself—it requires work to make it better.

Engineers analyze manufacturing variables during the early stages of process development. Improvement is difficult because few parts are available for analysis, meaning the low-volume production environment does not expose problems. Ideally, there is no guessing to determine variable settings, yet variables are often set based on "gut feeling" (which is a judgment call based on all previous technical experience).

Unfortunately, it is during high-volume manufacturing that previous minor process problems become major problems and create havoc for manufacturing. The best way to minimize this undesirable situation is with manufacturing team involvement in the early manufacturing process. Teams provide critical process know-how from a manufacturing perspective. This perspective is difficult for process developers to achieve because their emphasis is to meet a new product introduction date, not necessarily to produce product in high-volume manufacturing. Thus, manufacturing team involvement is essential for improvement throughout the product life cycle, from early process development through high-volume manufacturing.

7.2   LEVELS OF IMPROVEMENT

Effort expended in a process contributes to improvement. The three different ways a process can improve are:

- *Corrective maintenance*
- *Incremental improvement*
- *Quantum leap changes*

Corrective Maintenance

A process undergoes corrective maintenance to support its existing conditions across all process elements. There is no improvement relative to its current methods, only an effort to maintain the process. Corrective maintenance is the simplest effort to expend on a process, as it involves repairing equipment or reacting to defects after they occur, with no effort to improve the process by identifying and correcting the root cause of problems. *Corrective maintenance alone will not position a firm to be successful in a competitive market.*

It may seem that corrective maintenance is not improvement; however, if no minimal effort is done, then the process will degrade into ultimate disorder. Process degradation is a natural trend that occurs in any neglected process. Corrective maintenance effort prevents degradation, which is improvement relative to the degraded condition, but does not lead to process efficiency.

This minimalist approach to improvement is permissible if the marketplace requires no competitive products, which is basically what U.S. industries did during the period of capacity-driven production from the 1950s through the 1970s. Many manufacturing firms still only do corrective maintenance to support their equipment.

Incremental Improvement

The most efficient way to improve a process is through incremental improvement. This form of improvement analyzes existing process conditions, and then makes small process changes. Each incremental change is based on improving the least optimum condition of the current process. As any particular problem is corrected, then some other location in the process becomes the biggest problem, and is the new focal point for the next improvement effort.

Incremental improvement, also called **continual improvement,** is a long-term, team-based strategy that minimizes risk as the process undergoes change. This improvement is amenable toward teams because it is based on information about the existing process, with teams collecting and analyzing the information. Incremental improvement adapts as technology changes, eliminating preset limits as to how much a process can improve. For a manufacturing firm, incremental improvement is the surest method to achieve the competitive goals over time.

Quantum Leap Changes

The most difficult and risky form of effort for improvement is quantum leap changes, which involve a revolutionary change in technology and process that has little relationship to the existing process knowledge. An example of a quantum leap change is when the U.S. transportation system converted from horses to automobiles.

Firms should be cautious when using this form of change. It typically involves activities under research and development (R&D); however, quantum leap change is dangerous if this is the only improvement activity for a firm, because success is usually difficult to achieve. At the same time, some high-tech start-up firms are forced to develop their new product based on a quantum leap change, because this may be their only option and the high risk is understood.

7.2.1 Improvement Strategy

The most efficient manufacturing improvement method is incremental improvement. It builds on existing knowledge, uses small changes (evolution, not revolution), and is conducive to teams. It underscores the major problem while analysis occurs for the entire process. Because of this broad improvement effort, incremental improvement is the most difficult to implement and sustain. It requires in-depth process knowledge and a commitment against short-term problem solving.

Figure 7.1 shows how these three improvement levels interact in a company. It is best if the three types of improvement complement one another depending on the competitive situation of the product. In general, no firm should become too dependent on only one type of improvement effort. Unfortunately, some firms become complacent and depend excessively on corrective maintenance, or may even neglect maintenance altogether.

7.3 MANUFACTURING EQUIPMENT

In modern manufacturing, products are built through the use of equipment. Equipment (also known as tools or machines) is classified into three groups:

- *Automated equipment*
- *Semiautomated equipment*
- *Manual equipment*

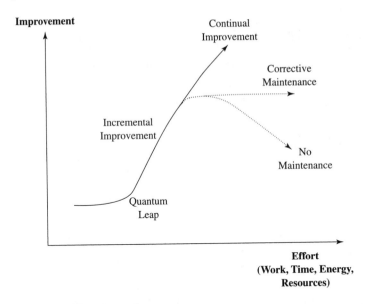

Figure 7.1 Improvement in Manufacturing

Automated Equipment

Automated equipment has been instrumental in permitting manufacturing technology to be controlled for more stringent product requirements. Automation requires minimal human intervention, primarily to interface with the equipment software and troubleshoot problems.

Product is automatically loaded and unloaded on automated equipment, with sensors to detect all critical conditions, such as position and time. Electronics interfaces with mechanical components (electromechanical systems), while software controls the tool and performs problem diagnostics (jammed part, missing materials, etc.). This high level of equipment integration with precise control of functions has contributed to standardization and repeatability in manufacturing.

The trend is toward more automated equipment that is software interfaced and linked through intelligent handling systems with other equipment, which leads to seamless equipment operations from tool to tool. Examples of automated equipment are the cluster tool and the track tool used in wafer fabs (discussed in Chapter 5).

Semiautomated Equipment

Semiautomated equipment requires additional human intervention, including operator loading and unloading of the product, setting of tool parameters (speed, voltage level, temperature, etc.), and operator intervention to diagnose problems. Many stand-alone tools (equipment without system integration) from the 1970s and 1980s were semiautomated, and are still in use in factories today. A comparison of the product flow between semiautomated and automated tools is shown in Figure 7.2, with the automated tool effectively controlling WIP on the conveyor with the use of sensors.

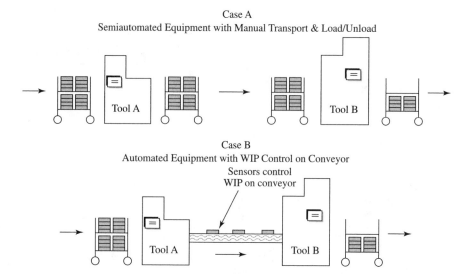

Figure 7.2 Equipment Integration for Seamless Flow

Manual Equipment

An operator performs all essential functions for manual equipment, with jigs or fixtures used for assistance. The fixtures assist small functions, such as part alignment and positioning. Parts are manually loaded and unloaded, and are usually transported between operations by the operator.

The operator uses judgment to set machine parameters and process parts on manual equipment. Having judgment and skill for manual equipment is an art that takes a long time to develop; however, the technical knowledge required to run and service the equipment is less than for automated equipment.

7.3.1 Equipment Strategy

It is not necessary to have only automated equipment in manufacturing. Each of the three tool classes is valid depending on the type of production (e.g., a manual tool that is used infrequently for a low-volume line, or automated equipment for a high-volume line with a changing product mix). Improvement efforts are needed regardless of the **equipment strategy;** however, to successfully compete in today's market, firms must be capable of automated, high-volume manufacturing that meets the three goals for competitive manufacturing.

If a process is continually improved, including suboptimal equipment conditions, then we eventually reach a state where it is necessary to advance the equipment. In this case, further improvement of the other process elements is limited by the suboptimal equipment, which may require a new level of automation to improve.

An automated tool integrates multiple process steps for increased standardization and repeatability.
Photo courtesy of Sematech Archives.

Automated equipment performance is critical in high-volume manufacturing because of the high cost to purchase, install, and operate the equipment, and the total dependency of the process on equipment performance. If automated equipment is unreliable and breaks down during production, then manufacturing cannot produce the product.

7.4 EQUIPMENT PERFORMANCE

Reliable equipment is important for high-volume manufacturing. It is especially critical for automated equipment with integrated, product-oriented flow, as there is little opportunity to modify the product flow if the equipment fails. Equipment failure thus stops the line with no throughput.

Examples of manufacturing problems resulting from poor equipment performance are:

- Frequent occurrence of product defects
- Excessive equipment downtime with poor product throughput
- Maintenance and engineering time spent responding to equipment failures
- Additional equipment capacity to cover equipment that is down

CASE STUDY

Automation

A cleaning station uses an automated tool to clean wafers in hydrofluoric (HF) acid. The current tool automatically loads wafers from a cassette into the cleaning station and exposes them to an HF cleaning cycle. Each cassette lot of twenty-five wafers must be manually put into the computer database prior to cleaning, and once again after cleaning is complete.

Wafers are sensitive to cleanliness, and a particle counter will count the average number of particles remaining on the wafer surface after a run. The wafer-surface particle count during a run on the existing tool averages 200 particles of 0.3 micron or larger, and often higher than the specification limit of 425 particles.

A new cleaning tool has been purchased and installed next to the existing tool. It has an increased throughput of forty lots of wafers in a 24-hour period. The new tool has automatic lot-scanning features based on bar codes, and determines whether the correct HF concentration is in the cleaning station for a particular lot. If a lot is loaded for the wrong HF concentration, then the equipment stops and a warning message flashes on the screen. The new tool leaves an average of only 50 particles on the wafer surface after a run, permitting a reduction in the specification limit to a maximum of 100 surface particles.

Points to consider

1. Given these improvements, and considering the new tool costs $1.5M, do you think the new tool should have been purchased? Why or why not?
2. What is the benefit of having the tool software confirm the correct product when the operator can do this?
3. What is the significance of reducing the specification because of the improved tool?

- Lack of equipment flexibility since equipment is always down, with no manual backup process available.

Equipment performance is described by an equipment reliability curve, sometimes referred to as the **bathtub curve** (see Figure 7.3). Initial equipment usage typically involves early failures, as seen in Phase I. These early equipment failures are usually due to defects escaping from the equipment build process, and are often found and resolved quickly (within 90 to 180 days). Examples of these defects are improperly crimped cable connectors or defective solder joints on circuit boards. Because new equipment is under warranty, this failure information is available to the equipment

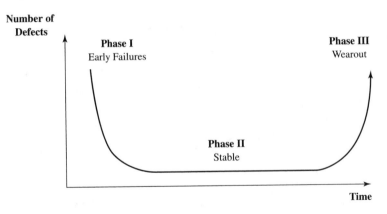

Figure 7.3 Bathtub Curve for Equipment Performance

manufacturers. It should be used by the manufacturers to implement corrective action plans for continual improvement. This effort will eventually reduce Phase I failures to zero.

With proper equipment maintenance, there is a stable period (shown in Phase II) in which equipment performance is good and should last for years. At some point, equipment enters the wearout period, Phase III, when performance problems increase significantly.

There are significant cost benefits to manufacturing if the stable phase can be lengthened. Improved equipment maintenance helps to accomplish this by moving the wearout phase farther out in time. Extending the stable phase lowers manufacturing costs by reducing the need to expend capital to purchase replacement equipment. It also introduces more flexibility for manufacturing production as to when the equipment must be replaced. This could be a significant cost benefit to a firm if technology is rapidly changing, as it permits the firm to purchase the most advanced equipment when it is available.

7.4.1 Chronic and Sporadic Losses

We achieve optimum equipment performance by understanding and correcting the reasons for equipment losses during production. To do this, we need to know the type of support and maintenance required for equipment improvement to extend the Phase II stable period. There are two types of equipment losses in manufacturing:[1]

Chronic Losses: Small, frequent equipment inefficiencies or failures regularly occur during production. These inefficiencies are perceived as a nuisance, yet are accepted as part of the equipment operation. They often grow

---

[1] See Reference 1 in the bibliography.

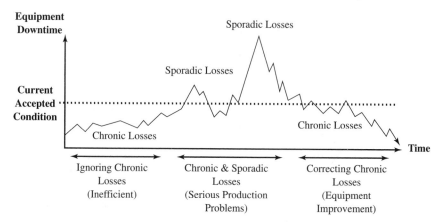

Figure 7.4 Chronic and Sporadic Equipment Loss

into major equipment problems (sporadic losses). With proper support and training, the operator or technician can do most corrective action. Chronic losses are hard to quantify and typically require innovation to fix.

Sporadic Losses: Major equipment failures that require a maintenance team with special equipment training for repair. Failures occur more frequently when chronic losses are extensive, because they push maintenance to "band-aid" equipment breakdown. Manufacturing is simply relieved to have equipment return to operation with *only* chronic losses. Effective equipment diagnostics and preventive maintenance (PM) reduce sporadic losses only if chronic losses are also reduced. Sporadic losses are generally corrected through equipment rebuild (e.g., replacement parts).

The relationship between chronic and sporadic equipment losses is shown in Figure 7.4. Note the **current accepted condition,** which is the acceptable ongoing level of losses, because no equipment improvement activities are underway. In other words, manufacturing has learned to live with these chronic problems, accepting them as though they are normal. Unfortunately, accepting a suboptimum current condition limits the equipment performance to the existing condition, with no possibility of continual improvement.

The goal for equipment downtime is simple: eliminate it. The production team will reduce sporadic equipment breakdowns only by addressing the day-to-day chronic problems that limit equipment performance. They must be capable of effectively addressing both chronic and sporadic losses to improve performance.

Because chronic loss exhibits itself as minor problems and occurs frequently, the team at the workstation can best supply the innovation necessary to address this type of equipment problem. The people working on the equipment have a unique viewpoint to interpret and react to chronic losses (such as noticing a change in the

---

✳   **CASE STUDY**   ✳
Discipline

To build working semiconductor devices on a wafer, many process steps and materials are required. A particular process step in a wafer fab is applying a thin coat of liquid polyimide (plastic material) to the wafer surface, followed by additional chemical processes. This operation is done on an automated tool known as a track tool, which has a load station for inputting multiple cassettes of twenty-five wafers, conveyors, and robot arms to move wafers between different process stations on the tool. Hold-down chucks hold the wafer in place during processing. To apply the liquid polyimide, a robot arm moves the wafer from the conveyor to a chuck and then a nozzle dispenses the liquid, followed by an oven cure and then additional chemical processing.

The polyimide track machine in this fab is not maintained properly for cleanliness. It has an excessive buildup of polyimide and other chemicals on the conveyors and chucks. Wafers are held unevenly during the polyimide dispense step, causing nonuniform polyimide thickness. The written machine-cleaning procedures require that the machine be cleaned daily, but workers ignore the procedure because they are busy.

Ten test sites on two wafers are measured for polyimide thickness from every two cassettes. At least one wafer typically fails the thickness requirement (too thin or too thick). When engineering is notified about the failure, the engineer usually signs off the lot with no further investigation.

Points to consider

1. Why is the machine-cleaning procedure not followed?
2. Who is responsible for making sure the cleaning procedure is adhered to in production?
3. What scenario would permit the operators to have more time if they were to follow the cleaning procedure?
4. Does it make sense to have a polyimide thickness requirement if parts that do not meet it are approved anyhow?

---

regular sound that equipment makes during operation, or seeking a way to stop inputting multiple data sheets for a new run). A disciplined workstation environment based on the 5S principles (CANDOS) will promote the detail-oriented problem-solving steps necessary to correct chronic losses. Chronic losses can often be repaired with minimal intervention from people outside of the process.

Examples of different types of chronic losses are:

- Adding extra time to an etch tool run to remove etching edge defects (instead of solving the root cause of why the defects occur)
- Filling in multiple data log sheets when one log suffices
- Long equipment setup times due to a disorganized work area
- Inspecting parts after each run to find defects instead of corrective action to resolve the root cause
- Early wear on a linear slide mechanism because of inadequate lubrication

On the other hand, sporadic losses result from major equipment problems and require an ongoing effort for preventive maintenance (PM) and early intervention techniques (such as data collection and analysis to determine when to do preventive maintenance). The special skills and equipment that are required for repairing sporadic equipment losses make these problems best suited for a maintenance team.

Improvement of chronic and sporadic loss necessitates constructive team actions to reduce the current accepted condition and predict maintenance repairs through effective preventive maintenance. It is unlikely an operator acting alone can attain the innovation required to correct chronic losses on modern automated equipment. At the same time, the maintenance effort on sporadic losses (major failures) can only improve equipment performance back to the current accepted condition. The team must work together to reduce chronic losses, improve the current accepted condition, and optimize equipment performance.

7.5 SOURCES OF POOR EQUIPMENT PERFORMANCE

Manufacturing costs from poor equipment performance can total a greater expenditure than the original purchase price of the equipment. In fact, if equipment is down or repeatedly producing bad product, then this performance can lead to catastrophic consequences for a firm operating in a competitive market.

Poor equipment performance comes from six major sources:

- *Poor design and build*
- *Incorrect installation and physical environment*
- *Improper maintenance*
- *Improper operation*
- *Nonimproving equipment measurements*
- *Politicized equipment performance*

✳ **CASE STUDY** ✳

Improving Maintenance Downtime

An older-generation sputtering tool in a wafer fab is used to deposit thin layers of a material on the wafer surface. Sputtering is a physical process during which atoms of argon gas are introduced in a vacuum system and ionized to a positive charge. The argon ions are accelerated to strike a grounded target material at high force, knocking off some target atoms and causing them to scatter. Some of the target atoms deposit as a thin film on wafers positioned in the vacuum chamber.

Two principal components located inside the sputtering tool are the target material and the shields, which direct the dislodged target material toward the wafers. The targets and shield must be replaced every 2 to 3 weeks after continuous use. This wafer fab has approximately six sputtering tools of this type throughout the process (it is an older-generation model).

The average time required for an equipment technician to change the target and shield on this sputtering tool was approximately 45 hours. During this time, the tool was not available for production.

A team of technicians evaluated the procedure to change the target and shield. It was improved by having all tools available and organized in a kit specific for this task. An adequate supply of targets and shields was held in stock, along with other components usually replaced during the procedure. The steps required for the procedure were documented and technicians were trained. By implementing these improvements, the average time to replace the target and shield was reduced to 12 hours (73 percent improvement).

Points to consider

1. Do you think the trade-off of having a team discuss this problem and the time spent organizing the workplace and tools were worth the reduction in time to change the target and shield?
2. Was this improvement reducing chronic or sporadic loss?
3. How was product throughput affected by this improvement?
4. Do you believe more improvement is possible?
5. Is this a high-risk investment of resources for the firm?

Poor Design and Build

Equipment that is of poor design and build creates problems in manufacturing that are difficult to overcome. In the worst case, equipment is scrapped and purchased from another vendor. One way to avoid this problem is to verify equipment performance at the time of installation or prior to purchase. The equipment engineer and the manufacturing team can be a part of this analysis, providing input for the content and size of the acceptance verification tests. Ideally, equipment performance tests are part of the purchase agreement.

Equipment is sometimes built with poor quality materials or shoddy work. Typically, equipment with these types of problems is related to a specific vendor who has not put effort into improving its own manufacturing process. Poorly built equipment is difficult to improve once it is in manufacturing. If this is the case, then the team must use the eight-step improvement plan (discussed later) to improve the equipment.

Incorrect Installation and Physical Environment

Incorrect installation of equipment can have a significant effect on equipment performance. For instance, sensitive inspection equipment that is not vibration-isolated from the floor may give erratic readings if the floor vibrates due to a passing fork truck. Another example is if compressed air does not have the proper filter, which permits contaminants in the pneumatic system. In another case, if equipment is exposed to chemicals in an environment for which it is not designed, then the equipment will degrade and have poor performance.

Improper Maintenance

Equipment maintenance is fundamental to optimum equipment performance. Equipment will not achieve or maintain the stable period (Phase II) with improper maintenance.

Even given the consequences of poor equipment performance, some manufacturing firms ignore equipment maintenance to increase production in the short term. Consider the bathtub curve model for equipment performance in Figure 7.3. Improperly maintained equipment causes premature machine wear, leading to a shortened period of stable operation in Phase II, and reducing the time to achieve equipment wearout in Phase III.

Examples of poor equipment maintenance are:

- No preventive maintenance (PM), with equipment operating until it breaks (this creates an ongoing crisis from one equipment breakdown to the next)
- Irregular or incorrect preventive maintenance
- Incorrect or missing schematics and maintenance procedures
- Improperly filled out or nonexistent equipment logs, which means problems must be reinvestigated each time they occur

Improper Operation

Equipment operation can lead to poor performance. Improper operation can be a subtle source of poor performance if the method of operation is based on prior practice, and usually assumed to be correct (because it has always been done this way).

Examples of poor equipment performance from operation are:

- Insufficient training or lack of training
- Unnecessary stoppages (lunch, break, etc.)
- Lack of operator discipline or inattentive work
- Improper equipment setup and adjustment

Nonimproving Equipment Measurements

Traditional manufacturing measurements for equipment are used to see if the tools perform acceptably. If these measurements are not reduced over time, then there is no equipment improvement. Nonimproving measures are equivalent to accepting the current condition with chronic losses as the best performance possible from the equipment. This acceptance blocks continual improvement, leaving the firm with suboptimum equipment performance and unable to compete in a competitive market.

The three traditional equipment performance measurements are:

Availability: Amount of time equipment is available to production (higher is better). It is usually expressed as a percentage.

Mean time to repair (MTTR): Average time it takes to repair a tool (shorter is better). It is usually expressed in hours.

Mean time between failures (MTBF): Average time a tool runs between breakdowns (longer is better). It is usually expressed in hours.

 EXAMPLE 7.1   AVAILABILITY, MTTR AND MTBF

A 1980s-era ion implanter for putting dopant ions into a wafer has gone down for equipment problems at the following times during the past month. The length of the repair time is also listed. Calculate the equipment availability, MTTR and the MTBF.

Incident 1	Monday	11/3/97	8:00 A.M.–9:00 A.M.	(1 hour)
Incident 2	Thursday	11/6/97	5:30 P.M.–8:30 P.M.	(3 hours)
Incident 3	Sunday	11/9/97	8:30 A.M.–1:00 P.M.	(4.5 hours)
Incident 4	Friday	11/14/97	2:30 P.M.–3:00 P.M.	(0.5 hours)
Incident 5	Thursday	11/20/97	3:00 A.M.–7:00 A.M.	(4 hours)

Solution

First, calculate the availability by adding the total downtime at 13 hours. Assume there are 720 hours in one month (30 days of 24 hours each).

$$\text{Availability (\%)} = \frac{(\text{Total Time} - \text{Total Downtime})}{\text{Total Time}} \times 100$$

$$= \frac{(720 \text{ hours} - 13 \text{ hours})}{720 \text{ hours}} \times 100$$

$$= 98.2\%$$

Second, calculate the mean time to repair (MTTR) by adding all downtime and dividing by the number of time it was down. For this equipment,

$$\text{MTTR} = \frac{\text{Total Downtime}}{\text{\# of Times Down}} = \frac{13 \text{ hours}}{5} = 2.6 \text{ hours}$$

Third, calculate the mean time between failures (MTBF) by adding the time interval between each incident and dividing by the number of intervals (four in this example).

Time between intervals 1 and 2: 80.5 hours
2 and 3: 60.0 hours
3 and 4: 121.5 hours
4 and 5: 132.0 hours

Total Interval Time = (80.5 + 60.0 + 121.5 + 132.0) = 394 hours

$$\text{MTBF} = \frac{\text{Total Interval Time}}{\text{\# of Intervals}} = \frac{394 \text{ hours}}{4} = 98.5 \text{ hours}$$

This equipment has an availability of 98%. The mean (or average) time it takes to repair this tool is 2.6 hours, and it goes down (on the average) every 98.5 hours, or approximately every 4 days. This availability appears good, but the MTTR and MTBF indicate less-than-desirable tool performance should only be used to document the existing tool performance. These targets should be reduced to motivate improvement.

It is acceptable to use these numbers for monitoring equipment performance as long as they are continuously reduced to reflect incremental improvement. Manufacturing should not become accustomed to a certain target and deem tool performance acceptable if it is met. If this occurs, then these measurements develop into the current accepted condition and are a hindrance to improvement.

To illustrate how misleading these equipment measurements can be in a complex manufacturing line, consider a tool availability target of 99 percent (the tool is available to production 99 percent of the time). This percentage is typically considered a reasonable availability target. The problem is that if a machine achieves this target, then frequently no additional effort is made to improve the machine. It is working as expected, so the "system" now accepts that the tool is not available to production 1 percent of the time.

If a manufacturing line has many pieces of equipment, even a 1 percent availability can create problems. For a 24-hour operation, as in most wafer fabs, 1 percent availability translates into about 1 hour of downtime every 4 days (because 4 days represents 96 hours of production). If there are 100 pieces of equipment, all with 99 percent availability, then all tools are randomly going down 1 hour every 4 days (on the average).

 EXAMPLE 7.2   TOOL AVAILABILITY

If there are 100 tools, each with an availability of 99 percent, what is the total line availability?

Solution

To solve, use the same approach demonstrated for obtaining the total process yield in Chapter 5 (see Figure 5.3). The availability is multiplied by itself 100 times. The simplest way to multiply a number by itself many times is to use the y^x key on a calculator. In this case, input 0.99 as y and 100 as x. The total line availability is:

$$\text{Total Line Availability (\%)} = (0.99)^{100} \times 100$$
$$\text{Total Line Availability (\%)} = 36.6\% = 37\%$$

This total line availability of 37 percent is the same as taking all 100 pieces of equipment in the process, and replacing them with only one piece of equipment that has 37 percent availability (assuming the line stops when a tool is down).

Based on Example 7.2, an apparently acceptable equipment availability of 99 percent is not desirable for process efficiency. This random 1 percent tool downtime occurring throughout the entire process drastically affects product flow by increasing the WIP waiting at each tool (parts wait at tools for the 1 hour of repair to be completed). This downtime increases product cycle time (which reduces throughput), requires additional maintenance support to troubleshoot and repair the equipment, and increases the risk of quality problems from equipment breakdown.

Politicized Equipment Performance

Automated equipment performance is critical to manufacturing efficiency. The production line stops if automated equipment is down. The addition of more people cannot alone solve this problem. The critical nature of automated manufacturing equipment means that poor equipment performance leads to politicized situations. Engineering, maintenance, and manufacturing expend their resources blaming each other and protecting themselves, instead of working as a team to resolve the problems. This environment is not conducive for continual improvement.

The unfortunate aspect of politicized equipment performance is that usually each area has some contribution to the poor performance. The creativity needed to improve subtle chronic losses along with ongoing equipment improvement requires a constructive team effort. Blame must be put aside so the team can find ways to solve the real problem—the equipment performance. Ultimately, this improvement is only accomplished through hard work, either as an individual with little organizational support, or as a team working together to assist one another to succeed against competitors. The best probability for success is the team approach.

7.6 OVERALL EQUIPMENT EFFECTIVENESS

To track team efforts for process and equipment performance, some manufacturers use **overall equipment effectiveness (OEE)**.[2] OEE is a diagnostic tool for analyzing process performance and assists in identifying problem areas. It uses four parameters in an overall equation to measure the fundamental aspects of equipment effectiveness: availability, operating efficiency, rate efficiency, and rate of quality.

The OEE measurement equation is:

$$OEE = Availability \times Operating\ Efficiency \times Rate\ Efficiency \times Rate\ of\ Quality$$

The team evaluates each of these parameters for the equipment and process under evaluation, and then calculates the OEE measurement equation. It is important that the team obtains accurate data that represent true equipment performance, or else all the effort is wasted. Often data are available on the workstation software database system.

Measuring these four parameters and summarizing them in the OEE equation help the team to assess process performance and identify areas that need improvement. To calculate each OEE parameter, use the following four parameter equations:

$$Availability = \frac{(Total\ Time - Total\ Downtime)}{Total\ Time}$$

Availability represents the time that the equipment is available for production. *Total time* is the number of hours in the measurement period, such as 168 hours in a week. Avoid using short time periods because it is difficult to accurately assess the performance. *Total downtime* includes all reasons the equipment is not available for production: equipment breakdowns, idle time for any reason, setup times, product qualifications, repairs, and maintenance (scheduled or unscheduled). Scheduled maintenance is the only legitimate reason for equipment downtime.

$$Operating\ Efficiency = \frac{Equipment\ Utilization}{Available\ Hours}$$

---

[2] See Reference 11 in the bibliography.

Operating efficiency measures how effectively the equipment is operated when it is available to production. Typical reasons for inefficient operation include no product to run or no available operator. *Equipment utilization* is the time the equipment is used for any reason, including regular product processing, engineering build requests, and test builds (such as processing a monitor part to control equipment parameters). *Available hours* is the total time minus the total downtime.

$$\text{Rate Efficiency} = \frac{\text{Actual Rate}}{\text{Theoretical Rate}}$$

Rate efficiency reflects the equipment's efficiency for producing product during operation. Poor rate efficiency could result from insufficient quantity of parts during a run (a partial load), rework of parts, or idle time due to no operator, lunch breaks, or team meetings. *Actual rate* is the quantity of parts produced during the equipment utilization time period previously defined, and can be extracted directly from the floor control software database at the tool. *Theoretical rate* is the number of parts produced if the equipment is processing parts under ideal conditions. Note that these rates also include load and unload times.

$$\text{Rate of Quality} = \frac{\text{Good Output}}{\text{Total Input}}$$

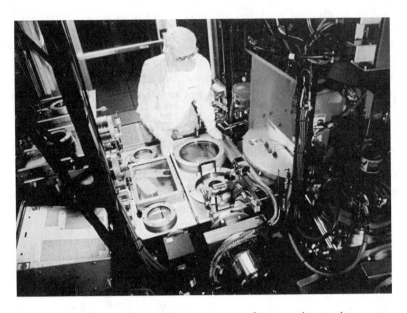

The team OEE analysis will evaluate all parameters of a complex tool to assist in problem identification.
Photo courtesy of Sematech Archives.

Table 7.1 Reasons for Major OEE Losses

Availability Losses: Reflects why the tool is not available to production	• Unscheduled downtime: breakdowns, repairs • Planned downtime: preventive maintenance • Tool setup • Product qualifications
Operating Efficiency Losses: Reflects nonoperation of available equipment	• No product to run • No operator available
Rate Efficiency Losses: Reflects product throughput of operating equipment	• Partial loads in tool • Equipment wearout • No throughput due to lunch, breaks, production team meetings, etc.
Rate of Quality Losses: Reflects the equipment's good output	• Process defects creating rework • Scrap parts

Rate of quality represents how good the product is. Any product that conforms to the product specifications with no rework or scrap is good output. *Good output* is defined as the total number of acceptable parts produced by the equipment and sent to the customer. *Total input* is the total number of parts that were started on the equipment.

To collect the OEE data, the team should measure data in the following order:

1. Availability: Measure the time the equipment is available to production.
2. Operating Efficiency: Measure the time the equipment is actually operated (when available).
3. Rate Efficiency: Count the number of actual product through the equipment (when operated).
4. Rate of Quality: Count the good product built on the equipment (when there is throughput).

A summary of the reasons for major OEE losses is shown in Table 7.1.

7.6.1 Calculating and Using OEE

When using OEE as a team tool, it is important to use reliable, consistent data that reflect the actual performance of the equipment. Data gathering may involve recording numbers with a stopwatch while observing the process, or collecting information from the floor-control computer database at the workstation. Make sure that the data are collected over adequate time to encompass all equipment variations. Additional information on how to collect manufacturing data is provided in Chapter 9.

✳ EXAMPLE 7.3 CALCULATING OEE

While processing wafers at a workstation, the team has analyzed equipment performance over several weeks (including all shifts) to determine the average conditions provided below. Calculate and interpret the OEE.

Sample operating conditions at a workstation

Total hours per week:	168 hours
Average downtime hours per week:	21 hours
Available hours of equipment:	147 hours
Equipment utilization per week:	112 hours
Theoretical rate of product:	50 wafers/hour
Actual rate of product:	30 wafers/hour
Total product started per week:	4550 wafers
Good product produced per week:	4025 wafers

Solution

These conditions are used to calculate the value (in percent) for the four individual parameter equations:

$$\text{Availability (\%)} = \left(\frac{\text{Total Time} - \text{Downtime}}{\text{Total Time}} \right) \times 100$$

$$\text{Availability (\%)} = \left(\frac{168 \text{ hours} - 21 \text{ hours}}{168 \text{ hours}} \right) \times 100$$

$$\text{Availability (\%)} = 87.5\%$$

$$\text{Operating Efficiency (\%)} = \left(\frac{\text{Equipment Utilization}}{\text{Available Hours}} \right) \times 100$$

$$\text{Operating Efficiency (\%)} = \left(\frac{112 \text{ hours}}{147 \text{ hours}} \right) \times 100$$

$$\text{Operating Efficiency (\%)} = 76.2\%$$

$$\text{Rate Efficiency (\%)} = \left(\frac{\text{Actual Rate}}{\text{Theoretical Rate}} \right) \times 100$$

$$\text{Rate Efficiency (\%)} = \left(\frac{30 \text{ wafers per hour}}{50 \text{ wafers per hour}} \right) \times 100$$

$$\text{Rate Efficiency (\%)} = 60.0\%$$

$$\text{Rate of Quality (\%)} = \left(\frac{\text{Good Product Output}}{\text{Total Product Input}}\right) \times 100$$

$$\text{Rate of Quality (\%)} = \left(\frac{4025 \text{ wafers}}{4550 \text{ wafers}}\right) \times 100$$

$$\text{Rate of Quality (\%)} = 88.5\%$$

The results from these four parameter equations are used to calculate OEE:

$$\text{OEE (\%)} = (\text{Availability} \times \text{Operating Efficiency} \times \text{Rate Efficiency} \\ \times \text{Rate of Quality}) \times 100$$

$$\text{OEE (\%)} = (0.875 \times 0.762 \times 0.60 \times 0.885) \times 100$$

$$\text{OEE (\%)} = 35\%$$

The overall equipment effectiveness resulting from these four equipment parameters is 35 percent. In other words, it took 100 percent of the time to produce what could have been produced in 35 percent of the same time period. This means the equipment is only effective 35 percent of the time to actually produce good product, with the other 65 percent consists of wasted effort attributed to the various OEE losses given in Table 7.1. These losses then show up in the operating conditions measured by the team and presented at the beginning of this example.

The team must now analyze each of these equipment performance losses to make improvements to the overall process, and give special focus to the cause with the lowest parameter (such as rate efficiency in the above example). It is of little benefit to focus on the highest parameter, because it will not dramatically improve OEE.

From the team's perspective, it is important to address all equipment performance conditions, because improvement in only one parameter will not drastically improve the equipment. For instance, in Example 7.3, if the team decides to work on rate of quality and improves it to 99.5 percent, this improves the OEE to only 40 percent. To make significant improvements in the OEE, *all four parameters must be improved, with special emphasis on the lowest parameter.*

One of the most powerful aspects of OEE is its ability to properly focus team resources on improvement efforts by highlighting areas with low OEE values. There is no rationale for teams to maximize their particular OEE at every operation in a process—it would be a waste of resource. The goal is for the operation with the lowest OEE to improve, thus improving the total process.

We learned in Chapter 5 that the operation with the lowest throughput is the process bottleneck (or constraint), because manufacturing can never work faster than the slowest operation. To optimize a total manufacturing line, focus resources on the bottleneck operation, because this will increase product throughput. At the same time, manufacturing cannot ignore nonconstraint operations, because the location of the constraint is dynamic and always changing in a process undergoing improvement.

The same situation occurs with OEE. To improve OEE, focus on the operation with the lowest value. As OEE improves at any operation, a new location in the process eventually becomes suboptimal with the lowest OEE. Teams must therefore work together to continually analyze and improve their process, preparing for the time when their operation is suboptimal. If teams ignore improvement at their operation for reasons such as, "We don't have the lowest OEE," then they will be caught unprepared when the situation changes, which is contrary to continual improvement.

 EXAMPLE 7.4   OEE AND CONSTRAINED MANUFACTURING[3]

An OEE analysis is being done for the three operations shown in the process of Figure 7.5.

1. If Operation 1 has 100 percent OEE, how much WIP is stored between Operations 1 and 2 after 1 hour?
2. How many parts are produced at Operation 3 after 1 hour? What is the OEE for Operation 3 (assuming OEE is defined by rate efficiency)?
3. Which operation is the process constraint? Where should be the focus for improving OEE? Is Operation 1 benefitting the process with 100 percent OEE?
4. Can you describe a situation where some stored WIP is acceptable?

Solution

1. The stored WIP between Operations 2 and 3 after 1 hour is three wafers, which is the balance of parts not used in Operation 2.
2. Assuming the equipment is available, is operating all its available hours, and creates no defects, then the OEE is defined by the rate efficiency. It can build

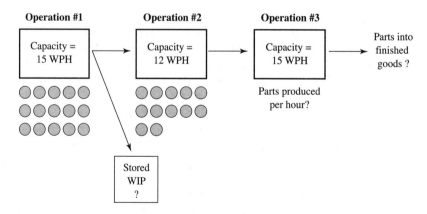

Figure 7.5 OEE and Constrained Manufacturing

[3] See Reference 11 in the bibliography.

fifteen WPH, but is only supplied twelve WPH, so after 1 hour there are twelve wafers produced. The OEE for Operation 3 is:

$$OEE\ (\%) = (12/15) \times 100 = 80\%$$

3. The process constraint is Operation 2, which should be the focus for improving overall OEE. There is no benefit for Operation 1 to have an OEE of 100 percent, and there should be a reallocation of resources to help Operation 2 reduce the bottleneck and balance the process.
4. Some stored WIP is acceptable to account for losses between Operation 1 and Operation 2, such as transport time and equipment downtime, but is only temporary, as the amount of WIP should be reduced as the process improves.

Although OEE can assist the team in identifying which aspect of an operation has priority for improvement, do not overlook problems with obvious solutions. These problems should be corrected first. If the OEE at a workstation needs to improve, then focus resources on parameters that are low relative to the others. Table 7.2 provides general guidelines for how to improve the four OEE parameters.[4] In addition, Appendix 6 provides a flowchart of team actions for low OEE.

As with any measurement, there is sometimes pressure to improve the measurement number without making fundamental improvements to the equipment or process.

Table 7.2 Process Considerations for OEE

Availability: All equipment should strive for high availability. Downtime reflects problems, and may impact product quality and yield. Preventive maintenance (PM) should be done on all tools, but focus efforts on constraint equipment. The goal is 100 percent availability (which may not be achievable due to PM time).

Operating Efficiency: The goal is 100 percent operating efficiency at the bottleneck operation. All other operations should be less than 100 percent, because downstream operations are not supplied sufficient work to be at 100 percent, and upstream operations only increase WIP if they are at 100 percent.

Rate Efficiency: The goal is again 100% rate efficiency at the bottleneck operations. Less than 100 percent rate efficiency is acceptable at nonconstraints, as long as it is not caused by tool problems. At nonconstraints, partial loads are acceptable.

Rate of Quality: Rework or scrap is never desirable. The goal is 100 percent rate of quality at all operations.

---

[4] See Reference 11 in the bibliography.

This is easily done through subtle data manipulation. Current estimates of equipment OEE in the electronics industry are approximately 30 to 35 percent. To avoid OEE becoming a marketing tool that publicizes maximum OEE numbers to outside parties or customers (without fundamental process improvement), OEE parameters should ideally be collected and used by the team to assist in the improvement process.

SUMMARY

Manufacturing improvement is the optimization of variables. There are three levels of improvement, with incremental improvement being the most effective for manufacturing. Reliable equipment performance is critical to modern manufacturing. The bathtub curve describes typical equipment performance. The goal is to prolong stable performance to extend the time to tool wearout, and is accomplished by team effort to reduce the current accepted conditions of the equipment (reducing chronic and sporadic losses).

There are six sources of poor equipment performance. OEE is a team tool to help identify where the team should start improving the equipment and process.

IMPORTANT TERMS

Manufacturing improvement	Poor design and build
Corrective maintenance	Incorrect installation
Incremental improvement	Improper maintenance
Continual improvement	Improper operation
Quantum leap changes	Nonimproving measures
Automated equipment	Availability
Semiautomated equipment	Mean time to repair (MTTR)
Manual equipment	Mean time between failure (MTBF)
Equipment strategy	Politicized equipment
Bathtub curve	Overall equipment effectiveness (OEE)
Chronic losses	Operating efficiency
Sporadic losses	Rate efficiency
Current accepted condition	Rate of quality

REVIEW QUESTIONS

1. List and discuss the three levels of improvement. Which method is most efficient?
2. What are the three types of manufacturing equipment?
3. Explain chronic and sporadic loss and the current accepted condition.
4. List the six sources of poor equipment performance.
5. Explain availability, MTTR, and MTBF.
6. State the OEE measurement equation, explain each parameter, and describe how this measurement can help a team allocate resources to improve.

EXERCISES

Sporadic and Chronic Equipment Losses

1. For the following equipment problems, list whether they are chronic or sporadic.

   - Machine breakdown due to lack of lubrication
   - Motor failure
   - Motor overload switch that must be reset regularly
   - Conveyor belt tension that must be readjusted each day
   - Lamp failure
   - Software glitch in controller
   - Incorrect fixture that requires intervention to manually align each run
   - Spray nozzle drips liquid and stains parts during runs

Equipment Measurements

2. A production tool has gone down for equipment repair during the following times over the past week. Based on the data, what is the tool availability, MTTR, and MTBF?

   Production Tool Downtimes

Monday	10:00 A.M. to 11:00 A.M.
Wednesday	5:30 P.M. to 7:00 P.M.
Thursday	4:00 A.M. to 5:30 A.M.
Saturday	3:00 P.M. to 6:00 P.M.

3. If you were confronted with a production line that had many tools with similar availability, MTTR, and MTBF as in the above problem, what would you recommend as a plan of action to improve the tool performance?

OEE

4. Your team is working in diffusion, and has estimated the following conditions for a furnace. Calculate the OEE for this operation. Based on the calculation, where would you recommend focusing improvement effort?

   Data for Furnace Operating Conditions

Total hours per week:	168 hours
Average downtime hours per week:	7 hours
Available hours of equipment:	161 hours
Equipment utilization per week:	105 hours
Theoretical rate of product:	40 wafers per hour
Actual rate of product:	30 wafers per hour
Total product started per week:	4550 wafers
Good product produced per week:	4375 wafers

8

CONTINUAL IMPROVEMENT

The difficulty for any improvement effort is maintaining the enthusiasm to improve. We have all done it—get excited about repairing a broken item, take it apart, only to quickly lose interest. We finish with a pile of parts and something that is in worse shape than before. It requires a focused effort to complete a project and improve over the previous condition. When this focused effort to improve is expended over and over, we have continual improvement.

OBJECTIVES

After studying the material in the chapter, you should be able to:

1. List each step and discuss the eight-step plan for continual improvement. Describe where iteration occurs in this improvement plan.
2. Describe the process platform model for controlling change during continual improvement.
3. Construct and interpret flowcharts.
4. Determine the root cause of a problem, including the use of a fishbone diagram.
5. Describe the actions necessary for implementing corrective action.
6. Construct a Pareto chart to prioritize problems.
7. Describe error-proofing in problem solving.
8. Explain cyclic and linear effort and how they contribute to continual improvement.

8.1 CONTINUAL IMPROVEMENT

Team improvement tools such as OEE can help us learn where to apply improvement effort, but now we have to learn *how to improve*. The question is, "How do teams really improve forever?" Haphazard improvement will not lead to continual improvement, and may even lead to process degradation. The team must be methodical and thorough to achieve the three goals: lowest cost, highest quality, and shortest delivery.

A structured approach to continual improvement in manufacturing is achieved with an iterative **eight-step improvement plan.** This approach is process based and encompasses all necessary actions for ongoing improvement. Because it iterates indefinitely, the eight-step plan is self-sustaining. When it is linked with statistical process control, it guides the team to continual improvement (to be explained in Chapter 11).

Note that the eight-step plan is for ongoing improvement, yet it encompasses many of the same elements necessary for troubleshooting a broken piece of equipment. An equipment technician or field representative of an equipment supplier usually does troubleshooting. Equipment troubleshooting may use a different order of events than those outlined in the eight-step plan, because the immediate goal is to repair the equipment and return it back on line to production. Ultimately, all equipment troubleshooting information should be input into the manufacturing team for use during continuous improvement.

The iterative eight-step improvement plan for continual improvement is:

- *Step 1: Use team approach*
- *Step 2: Analyze initial conditions*
- *Step 3: Collect and analyze data*
- *Step 4: Identify root causes of problems*
- *Step 5: Take corrective action*
- *Step 6: Use error-proofing*
- *Step 7: Implement process control and iterate*
- *Step 8: Continually improve*

Step 1: Use Team Approach

The team is the foundation of the optimization effort. If possible, seek team members with complementary skills and backgrounds. Teams may include the operators and technicians who work with the tool, the equipment engineer, the maintenance technician, and a team leader. Others are included depending on the nature of the problem.

Teams link operations across the process, providing a means of self-sufficiency for improvement activities. They create flexible, knowledgeable resources that can confidently solve problems while supporting continual improvement.

Step 2: Analyze Initial Conditions

The second step is critical in that it involves analyzing the process to establish the initial conditions. These initial conditions are rich with potential information about the nature of the problem. This information will be useful throughout the entire improvement effort to provide a basis to understand future conditions.

The initial analysis establishes the first body of knowledge about the equipment, called a **process platform.** We use the term *platform* because it signifies a distinct body of information about the operation undergoing analysis. This permits us to break our work into stages, simplifying the improvement effort at each platform. Each platform represents process conditions for how the product is built until a conscious effort is made to change the process.

Platforms are a basis for comprehending change in a dynamic process, merging present and future knowledge about an operation into incremental steps. It creates team confidence for ongoing improvements to a complex manufacturing line. Without the use of platforms to view the work, manufacturing runs the risk of making changes without really understanding where the process is located on the improvement path. The basic process platform model is shown in Figure 8.1.

When a team is on a platform, they make controlled changes to the process from this platform. They confidently return to the prior platform if unforeseen problems arise during the change implementation. Platforms are as much a state of mind as they are concrete information about the operation to be improved. We will use this model of process platforms throughout our improvement activities to provide a struc-

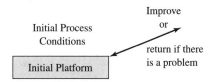

Figure 8.1 Initial Process Platform for Continual Improvement

ture for manufacturing improvement. Note that the term *process platform* is a model to understand improvement, and not commonly used in the practice of manufacturing.

Flowcharts   Another tool for establishing process conditions is the flowchart. Flowcharting describes a process from start to finish by defining each step and using arrows to show how the steps interact. A flowchart is effective as a team tool, and should be displayed so that all team members participate in its construction to help them understand the individual process steps and how the steps are related.

Flowcharts can involve elaborate symbols, but it is best to keep them simple with two basic symbols linked by arrows to show the direction of flow. The symbols are a rectangle for a task and a diamond for a decision, as shown in Figure 8.2. *Include all steps,* even those thought to be unimportant. An example of a flowchart is shown in Figure 8.3

A flowchart is conceivable for any process, large or small. It helps team members define the inputs and outputs to the process, visualizing flowloops (such as rework after inspection), and illustrates what is done to the product at each step. Knowing this information helps establish the initial process conditions, and may prove instrumental in solving problems.

Process Observation   Be cautious during this initial analysis stage. Carefully observe and document trends that appear after repetitive operation of the equipment and process. Avoid changes without understanding the key variables in the process. Haphazard changes (e.g., just hoping to find the solution) can make the problem worse, plus cover up key information about the problem.

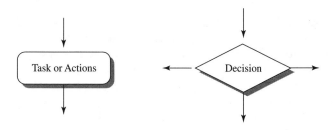

Figure 8.2 Flowchart Symbols

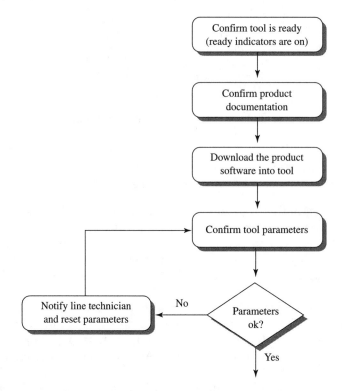

Figure 8.3 Flowchart for Downloading Tool Software

Typical activities during **process observation** are:

- Observe all aspects of the process, using experience and process knowledge as a basis for judgment. Try to identify key variables.
- Discuss the equipment operation and findings with other operators and team members.
- Document all conditions in a shift log or equipment log.
- Review documentation and understand differences in interpretation.

While establishing the initial process conditions, we sometimes need to implement a short-term solution to a problem, which is often necessary in manufacturing so that production can meet its scheduling commitments. It may be done with inspection, rework, or some other temporary actions and is stopped once the ultimate solution is implemented. It is usually a more costly solution than the optimum approach.

Manufacturing should not regularly use short-term solutions to solve production problems. If this is the case, then the process needs fundamental improvement (as described in this book). Always discuss short-term solutions among team members so that everyone understands how the solutions will be used, and what the cr long-term actions are. Everyone should be aware that short-term solution that—short term. Unfortunately, many firms permit short-term solutions to evo. into long-term "fixes," leading to serious problems for competitive manufacturing.

A final point is that during problem solving, it is important to correct obvious problems (such as finding an underrated fuse that makes a tool trip its circuit breaker). Never let problems go uncorrected if there is an obvious solution.

 EXAMPLE 8.1 TROUBLESHOOTING AN EQUIPMENT PROBLEM

A piece of automated equipment goes down during production. You are a new equipment technician, and unfamiliar with this tool. You usually work with another technician who has several years experience, but is in a training class today. List six options you have to troubleshoot this equipment, in order of priority to get it back on line.

Solution

1. Observe the problem on the equipment.
2. Talk with the operator about the problem.
3. Review the operating procedure to ensure that all steps have been followed.
4. Check the equipment log for a history of this problem or similar problems.
5. Talk with an equipment technician who has experience on this tool.
6. Review the equipment documentation.

Initial troubleshooting solutions deal with acquiring information about the equipment and problem to make intelligent decisions. The technician should not be removing parts, replacing components, or doing similar repair—this activity occurs only after the root cause of the problem is understood.

Step 3: Collect and Analyze Data

The team collects numerical data from key process variables to use for problem analysis. Proper data collection and analysis provide complementary insight into the information obtained during the initial investigation. Use information obtained from the initial investigation to assess which process variables are critical for data collection.

The tool available to the team for analyzing numerical data is statistical process control (SPC). This statistical analysis tool gives the team insight into process performance and provides a basis to predict future problems. SPC is covered in depth during the next three chapters.

The first time the team uses the eight-step plan, the data collection and analysis will be crude, which is understandable because the team is learning how to acquire information about the equipment and process. As the team iterates between steps 3 to 8, each cycle will refine the variables, how the data are collected, and the analysis of the data. The team becomes smarter.

As the team improves, information about process changes is documented through analytical tools such as the SPC chart history, team meetings, shift logs, equipment logs, and manufacturing procedures, and is available to the team during the decision-making process. Each change condition in the eight-step plan can thus be viewed as a process platform.

Always strive to obtain good numerical data (avoid GIGO: garbage in–garbage out). Collecting bad data and then using the facts in the decision-making process is waste and is potentially dangerous for the product. Simply put, *no data are better than bad data.*

Step 4: Identify Root Causes of Problems

Once the team has conducted an initial analysis of the process and is obtaining information about the nature of the problems, it is necessary to identify root causes. Identifying the root cause is finding the true source of the problem, not a symptom that only exhibits the problem. Consider a patient who is ill with a fever. The fever is a symptom, whereas the root cause of the fever must be diagnosed (such as having the flu). Treating the fever alone may not correct the root cause.

An operator collects data for problem analysis.
Photo courtesy of Ion Implant Services, Inc.

Finding a root cause of a problem involves analyzing all variables associated with the problem and how they interact. The focus here is finding the right process variables that cause the problem. Different problem-solving techniques available to the team are:

- Continually asking "why" each time a new aspect of the problem is uncovered
- Taking a break, sleeping on the problem, and developing a new perspective
- Using brainstorming with team members
- Looking for all possible root causes

Avoid haphazard investigations based on personal opinions. A methodical approach to finding root causes is important, because they are usually buried beneath superficial causes. If an investigation leads to the wrong root cause, then corrective action is taken on a cause that will not fix the problem. Needless to say, this is not good for improvement.

Brainstorming is the idea-generation method covered in Chapter 4, and is a powerful problem-solving tool. It brings out ideas from all team members, helps team members "feed" off of other ideas, and if used effectively, is a powerful team tool for finding root causes.

Fishbone Diagram   A fishbone diagram is a visual tool used to establish all potential causes of a problem. It is commonly used with brainstorming. This diagram looks at the problem (an effect) and lays out all the potential causes for the problem in a picture format (the causes). It is also known as a **cause-and-effect** diagram or an Ishikawa diagram (named after Professor Ishikawa of Japan who first used the technique).

The most effective way to lay out a fishbone diagram is by categorizing all potential causes by the four process elements: people, machines, methods, and materials. A fishbone diagram highlights this in Figure 8.4 for a manufacturing problem

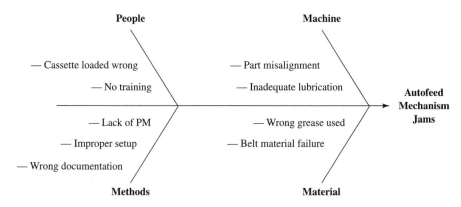

Figure 8.4 Fishbone Diagram for Autofeed Problem

with a load/unload autofeed tool (automatically feeds parts into equipment) that repeatedly jams.

A method to verify the correctness of a suspected root cause is the **on/off test.** The principle behind this test is that if a suspected cause is truly the root cause, then a controlled change in this cause should turn the problem on and off. If you have the true root cause, you can change the root cause and make the problem go away, or change it back and the problem returns. This may not be true in every case, depending on how the different variables interact with the problem, but it is a surprisingly effective test (especially when the team is under pressure to attain results).

Step 5: Take Corrective Action

Corrective action is a controlled change to a process to correct a problem. The team takes corrective action to resolve the problem once the root cause is identified. Avoid haphazard changes because this leads to confusion.

Change because of corrective action is inherently unstable for any process. It is like using rocks to cross a stream. While on shore, we are stable. While crossing the stream by stepping on the rocks, we risk slipping into the water. The act of changing is inherently risky. Once we reach the other shore, we are stable again.

This is why the process platform concept is important during the change period of taking corrective action. It breaks the work into smaller entities, building workers' confidence as they learn to understand a process, and providing a link between process conditions during the actual change activities. This relationship between change and process platforms as a basis for process improvement is shown in Figure 8.5.

The actions necessary for implementing corrective action to a process are:

- Obtain team consensus
- Set priorities

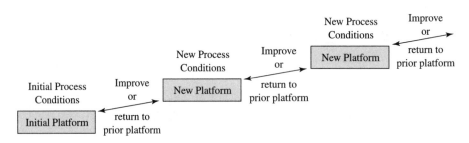

Figure 8.5 Process Platform Model for Process Improvement

- Attain approvals and document
- Implement a single change at a time
- Verify changes lead to improvement

Obtain Team Consensus Achieving team consensus is the most efficient method for making decisions about how to make incremental changes to a process. Team members should work together to review and interpret data collected from the process, and determine the best way to begin the change. Sometimes achieving consensus is difficult, while other problems are so complex that only a few technical people understand the solution. In most cases, effective team action is the shortest path to correct problem resolution.

Set Priorities Set priorities to focus team resources during change. The team may have a "gut feeling" for which problem has priority after the initial analysis, or there may be other problems that have obvious solutions. Serious problems are often logical starting points (for instance, a high product scrap rate).

A well-known procedure for prioritizing problems is the **Pareto chart** (shown in Figure 8.6). This chart categorizes the various problems found in a piece of equipment or process, and plots them in order of importance (such as on a chart for frequency of occurrence). This visual chart helps the team prioritize problems, permitting action on the larger problems first, and is important because solving big problems can move the process quickly toward improvement.

A Pareto analysis of problems is beneficial because typically only a few big problems are the root cause for most defects in manufacturing. It has been estimated

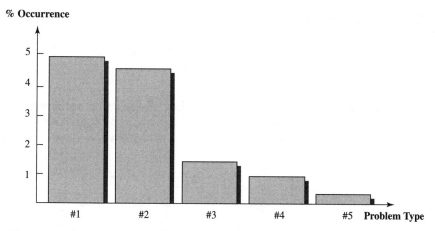

Figure 8.6 Sample Pareto Chart for Defects

that 20 percent of all problems lead to 80 percent of the product defects.[1] Big problems are sometimes easy to correct (for reasons such as they occur more frequently, can receive more attention in the organization, and may be more easily analyzed for trends).

Conversely, small problems are often difficult to identify, measure, and correct. For instance, if a process has a 50 percent yield, it may take a short while to correct the major yield detractors and improve to 95 percent yield. However, it could take a long time to improve to 98 percent, and an even more difficult effort to continue toward 99 percent, followed by 99.9 percent, and so on.

 EXAMPLE 8.2   CONSTRUCTING A PARETO CHART

Resist is a light-sensitive liquid material coated on the wafer surface and then cured so that it can be patterned during the photolithography operation. The resist is applied by spinning a wafer and then dropping a small amount of the liquid onto the spinning wafer. Excess resist is spun off while a thin coat of several microns thickness remains for further processing.

There is considerable operator time spent cleaning and working on the resist tool, and the team has analyzed these activities and found the following problems:

1. Excessive time spent cleaning the resist machine: 30 minutes/shift
2. Operator mishandling of conveyor, causing downtime: 5 minutes/shift
3. Contamination of equipment with resist: 35 minutes/shift
4. Replenishing rinse chemicals: 10 minutes/shift
5. Resist contamination getting on other surfaces: 20 minutes/shift

Draw a Pareto chart showing how these problems are related.

Solution

Use 720 minutes in a 12-hour shift, and calculate the percent occurrence for each type of problem. List all the problems on a chart in order of occurrence, with a bar graph to show the percent of time it occurs. A sample calculation is done for the most frequent problem.

$$\% \text{ Occurrence of \#3} = \frac{\text{Occurrence Time of \#3}}{\text{Total Time}} = \frac{35 \text{ minutes}}{720 \text{ minutes}} \times 100 = 4.9\%$$

---

[1] See Reference 18 in the bibliography.

% Occurrence

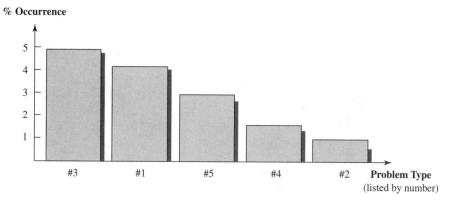

Pareto Chart for Spin Resist Problems

Obtain Approvals and Document All manufacturing changes must be approved and documented, which gives a broader perspective to the change in case there is an unanticipated problem or it affects another production area. This procedure also alerts documentation control to update the specifications. The team usually does not have the approval authority to implement changes. Each company has different procedures for implementing process changes, so ensure that all approvals are obtained.

The team documents changes in the operation log, including information such as date, time, lot number, and tool number. Documentation is important in case a problem arises due to the change, requiring a return to the previous process condition while the parts are traced and pulled out of production if necessary (this would be for a severe case, but be prepared for the worst).

Implement a Single Change Single-variable change is the most effective way to introduce process or equipment changes in manufacturing with minimal risk of failure. Single-variable change is conservative, permitting the team to control the change, observe the process after the change, and verify that the change has improved the problem. If a problem arises, return to the previous process platform to analyze what went wrong prior to making the next change.

Statistical purists will disagree with this approach. Their argument is that manufacturing must use a multivariable method to analyze processes. This method studies several variables simultaneously to assess their interaction. From a statistical standpoint, the multivariable method may be more desirable; but statistics is a tool for action to improve the process, not a goal within itself.

The goal here is achieving optimum manufacturing conditions. Multivariable evaluations become overly complex, putting more emphasis on the statistical method to the detriment of understanding the equipment and process. The eight-step improvement plan is based on solid knowledge of the process. Single-variable change is

Every process requires evaluation and error-proofing for improvement.
Photo courtesy of Sematech Archives.

efficient when used with knowledge acquired during the initial process investigation, and is retained through the use of process platforms.

Verify Changes   The team must verify that changes lead to improvement. There is always a risk that we do not properly understand the root cause to a problem under improvement. Verify improvement by closely monitoring data collected during the time of the change. If poorer process performance results during a change, then return to the previous process platform and analyze why the change was unsuccessful.

Step 6: Use Error-Proofing

Error-proofing is a method of ensuring that a corrected problem does not return at a later date. If the problem returns, then the team is not incrementally improving, but instead improving sporadically and at times even regressing. Continual improvement is unattainable in these conditions.

The concept of manufacturing error-proofing is from the Japanese term *Poka-Yoke,*[2] and involves implementation of simple equipment or process changes that will not permit the same problem to occur again. For instance, if a part can slip off the edge of a platform, then mechanical stops are mounted at the platform edge to prevent this problem. Modern error-proofing methods use electronic devices integrated

---

[2] See Reference 12 in the bibliography.

with the system software to ensure that the problem is corrected. Error-proofing can also involve simple techniques, such as an indicator placed on a flow-valve knob to show the correct position, known as a visual control.

Step 7: Implement Process Control and Iterate

If a process undergoes change for improvement, then process control is important to monitor the change and verify improvement occurs. Statistical process control (SPC) provides this information to the team through its powerful statistical prediction techniques. It continually monitors process performance as the team iterates through steps 3 to 7 of the eight-step improvement plan. While iterating, the team understands and corrects process problems, and accumulates knowledge represented in a process platform. As more data are collected and analyzed, more process knowledge is obtained, and the team becomes more in control of its operation.

The eight steps in the eight-step plan are repeated here to show where iteration occurs.

Step 1: Use team approach

Step 2: Analyze initial conditions

Step 3: Collect and analyze data

Step 4: Identify root causes of problems

Step 5: Take corrective action

Step 6: Use error-proofing

Step 7: Implement process control and iterate

Step 8: Continually improve

Statistical process control techniques verify stability and are used to move the process to the next platform for further improvement. We will learn in Chapter 11 how the team uses SPC to demonstrate stability and then force improvement to a new process platform, guiding the team through change and improvement.

Successful process improvement occurs as the team continuously iterates through the eight-step plan. This iteration cannot stop. A single or partial pass through the eight-step plan would be a wasted effort and will never move the process to continual improvement.

Cyclic and Linear Effort We now have sufficient knowledge to philosophically describe how the process platform model represents team efforts to continually improve. *There are two basic efforts a team must simultaneously expend for continual improvement:*

Cyclic effort: Continually iterate between steps 3 and 8 to resolve problems at an operation

Linear effort: Incrementally move the process from platform to platform in the direction of continual improvement

The **cyclic effort** from the eight-step plan represents the day-to-day improvement work done by team members at the workstation. It may involve specific actions such as correcting a faulty flow-valve or finding a simpler way to load a fixture in a tool. It represents the hard work, commitment, and discipline that sincere team members put into their process.

Linear effort is the long-term work necessary to uncover new, previously accepted problems. This work forces the process to a new condition and level of improvement. It guides teams for improvement, because each new process condition has its associated problems and new cyclic effort needed to correct them.

Statistical process control is the fundamental tool used with both of these efforts. We will learn in Chapter 11 how SPC is used by the team to accomplish cyclic and linear effort for continual improvement. The relationship between cyclic and linear effort is shown in Figure 8.7.

With our structured approach to continual improvement, the importance of these two simultaneous efforts is obvious. We cannot just work to resolve problems in a process (cyclic effort) and assume that continual improvement will follow. Nor can we talk about continual improvement (linear effort) while ignoring the hard work that must occur in the process prior to making the change.

If we only expend cyclic effort and iterate through the eight-step plan, we are not forced on a linear path to improvement. We work hard to fix daily problems, but our process does not improve, meaning we are responding to problems instead of performing ongoing improvement. Conversely, only expending linear effort to move from change to change (platform to platform) without understanding the process conditions prior to making changes means we move aimlessly. We never acquire sufficient knowledge to make a reasonable change because no work is done in the process.

At any instant in time, if considered in isolation, cyclic and linear efforts appear opposed to one another; however, with the process platform model, we know these two efforts are complementary. Understanding that we have two overlapping ef-

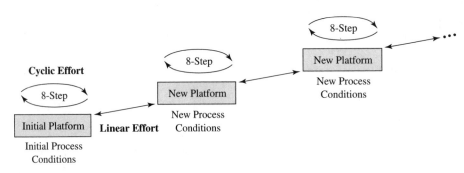

Figure 8.7 Model of Cyclic and Linear Effort for Continual Improvement

forts—a cyclic effort for acquiring knowledge and a linear effort to guide us toward perfection—provides the foundation for visionary team improvement.

Step 8: Continually Improve

As the team iterates through the eight-step plan, the process will continually achieve a new stable condition around a process platform, undergo analysis, change for the next improvement, and attain a new process platform. Statistical process control is the team tool used to guide the team while on the path of continual improvement. In this manner, the team understands the improvement process and is in control. If an unanticipated problem occurs during the change period and is not understood, the team resets the process to the previous platform, thus avoiding a major problem for production.

A process continually improves if the team expends cyclic and linear effort. Work in the process corrects problems (cyclic effort), and work to uncover previously accepted problems moves the process toward higher platforms of performance (linear effort).

The objective of continual improvement is not to know how well the process and equipment perform in absolute numbers, because that state will never be determined (i.e., the absolute zero condition of zero defects is not critical, only that the process is moving toward zero defects). The goal is incremental improvement in the proper direction. No one can state when any process becomes absolutely perfect—it is statistically impossible. However, merely the act of striving for perfection creates a foundation for team harmony and accomplishment.

When teams attain this state of harmony from improvement, they no longer have to "firefight" production problems. The structured improvement process is the glue that bonds the team together, molded as one into the process and equipment, improving efficiency, quality, and productivity toward competitive manufacturing.

The team now works together, in control of its destiny and achieving concrete results from continual improvement. As to when the team actually achieves the three competitive goals of lowest cost, highest quality, and shortest delivery, that can only be determined relative to the competitive market. Perfection is the sole objective of the team's state of mind, with a relentless desire for further improvement.

SUMMARY

Continual improvement occurs by use of an eight-step improvement plan, which employs basic improvement tools to iterate through problem analysis, data collection, root cause determination, corrective action, error-proofing, and process control. Initial process conditions are documented through observation and data collection. A process platform model for change is developed to categorize information during the change process. Flowcharting is a team improvement tool that illustrates interdependencies in a process. A fishbone diagram is helpful for analyzing root causes to problems, and a Pareto chart assists in prioritizing problems. Corrective action should be limited to a single change at a time, and requires approval and documentation. Error-proofing is important to ensure that problems do not return. The process platform model encompasses work in the process (cyclic effort) and work to move a process to a new level (linear effort). With the process platform model and continual iteration through the eight-step plan, a process moves toward continual improvement.

IMPORTANT TERMS

Eight-step improvement plan	On/off test
Process platform	Corrective action
Flowchart	Pareto chart
Process observation	Error-proofing
Root cause	Cyclic effort
Fishbone diagram	Linear effort
Cause-and-effect	Continual improvement

REVIEW QUESTIONS

1. State the eight-step improvement plan. Between what steps does the plan iterate?
2. Why is initial analysis of the problem important?
3. What is a process platform and how does it benefit manufacturing improvement?
4. Describe how to build a flowchart.
5. Why is process observation important during the initial analysis?
6. Why are data collected during process improvement?
7. What is a root cause, and why is this important for improvement?
8. Explain how a fishbone diagram assists in finding the cause for a problem.
9. List and explain the actions necessary to implement corrective action in a process.
10. Explain the Pareto chart and how it benefits corrective action.
11. What is error-proofing in a process?
12. Why is SPC important during change for improvement?

13. Explain cyclic and linear effort and how they represent work to improve a process.

14. Describe how the eight-step plan leads to continual improvement in a process.

EXERCISES

Troubleshooting under Pressure

1. You are troubleshooting the equipment problem in Example 8.1. There is pressure to quickly diagnose the cause of an equipment problem that is down during production. In addition to the six problem-solving approaches given in the example, list at least six additional sources of information (people, documents, etc.) for quickly solving this problem. Be specific and prioritize them based on quickly getting the tool back on line.

Flowcharts

2. Make a flowchart showing all the process steps required to change a flat tire on a car.

3. Refer back to the process flow diagram in Figure 1.1 in Chapter 1 for building wooden cabinets. You are an experienced table saw operator, and want to show all the steps to new operators for setting up a table saw with a positioning stop. The major steps are:

   • Put positioning stop at correct location (with a tape measure) and bolt it down.
   • Adjust height of saw blade.
   • Make first cut with setup board.
   • If setup board length measures to spec, then proceed with production lumber.
   • If setup board length is not in specification, then cut another setup board and recheck dimension.
   • If setup board continues to measure incorrectly, then notify senior operator.

   A. Draw a flowchart that shows these steps and how they are interrelated.

Fishbone Diagrams and Root Causes

4. Use a fishbone diagram to find all potential causes for one of the following problems:

   • Car will not start.
   • Stereo does not turn on.

- Bicycle chain keeps coming off.
- Room light does not turn on.

Once you have identified all the potential causes of each problem, describe how you could use the on/off test to know if a cause is a true root cause.

5. Review the following defects that occur during the test of wafers at the wafer sort operation. The frequency of occurrence over the past month is listed. Create a fishbone diagram and categorize each defect under one of the manufacturing elements.

Table of Defects from Wafer Sort (Test)

Defect	*Frequency*
Glass on contact needle (probe needle cannot contact pad)	18
Open circuit inside die	12
Alignment target on wafer too dim to read for positioning	8
Cables connecting probe card to tester are loose	5
Lack of tester PM for alignment	2
Bent probe card needles	28
Incorrect alignment of first die, affecting all follow-on die	9
Incorrect parameters in test specification	1
Improper probe card installed in tester	3
Vacuum chuck holding wafer is not adjusted properly	8
Incorrect pressure on probe card pins contacting pads	13
Incorrect wafer software loaded into machine	5
Total Defects	112

Pareto Chart

6. Use the defect information given in Exercise 5 concerning wafer sort, and construct a Pareto chart. Which problems warrant immediate attention? Which problems do not occur often, but may be easy to correct?

Error-Proofing

7. List six examples of error-proofing for common events. An example is the requirement to push in the clutch pedal of a four-speed transmission prior to starting a car (or putting an automatic transmission into park).

8. For each of the following manufacturing situations, give an example of how you might error-proof the action to prevent a problem:

   A. A liquid flow-valve must be opened a certain number of turns for each different product part number processed in a tool.

   B. A bolt continually loosens on a tool and risks falling out if it is not tightened.

   C. The run lot number is sometimes typed incorrectly into the computer during log-in before a run.

   D. The conveyor is manually adjusted for different product sizes, with a potential of a conveyor jam if it is misaligned.

9

BASIC STATISTICS FOR IMPROVEMENT

The average person is intimidated by statistics. If you ask most people the probability of getting heads from a coin flip, however, they usually know the answer because they have a basic knowledge of statistics. Most people need insight to understand and use statistics in practical ways such as for improvement.

Some sources of improvement are obvious and require little statistics, such as using a doorstop to hold a door open while unloading furniture. Other sources of improvement are not so obvious, such as finding the root cause of an intermittent tool problem that creates product defects. Statistics helps to identify and solve complex problems by clarifying the relationship between variables.

Manufacturing improvement involves a combination of the obvious and not so obvious. Manufacturing personnel must have the ability to respond and correct the obvious problems, and possess the analytical tools necessary to correct complicated problems. Statistics is the tool needed to analyze different manufacturing variables and assess their importance in solving complicated problems.

OBJECTIVES

After studying the material in this chapter, you should be able to:

1. Define *statistics* and explain why it is important in manufacturing improvement.
2. List, define, and give examples for the two types of data.
3. List and explain the two types of variation.
4. Explain when a process is in control and out of control.
5. Calculate the probability of an event based on possible outcomes.
6. Define *manufacturing data* and how they are used in SPC, and list the five requirements for collecting good data.
7. Describe a histogram, and be capable of constructing and interpreting a histogram from raw data. Define and explain: *nominal dimension, USL, LSL,* and *process window.*
8. Define *central tendency* and the *mean*. Calculate the mean for a group of data.
9. Define *variability, range, standard deviation,* and *sigma*. Calculate range and sigma for a group of data.
10. State what two parameters define the normal curve, and how sigma relates to the probability of occurrence in a normal distribution.
11. Given the mean and sigma for a population, calculate the probability of occurrence and the number of parts outside a specified limit (using the Z statistic when necessary).

9.1 WHAT IS STATISTICS?

Statistics is defined as the collection, analysis, and interpretation of numerical data that represent a **population.** The importance of statistics is based on its ability to determine the relationship between variables, and to use a sample population to make predictions about a larger population. For example, political pollsters survey the political beliefs of a small group of people (around 1200 voters) and statistically analyze the data representing their beliefs to make predictions about the U.S. voting public (close to 100 million people). If a pollster had to poll all voters to see how they would vote, it would be costly and time consuming, and perhaps impossible.

In manufacturing, statistics is used for three primary reasons:

1. **Process Control:** Collect and analyze data to give information about a process and its performance.
2. **Experimentation:** Obtain numerical data through data collection and analysis to understand variables.
3. **Test:** Product measurements to ensure that work is acceptable.

Our studies will focus on using statistics for process control by measuring equipment and process variable conditions and statistically analyzing the data. By measuring and analyzing data about variables in the process, a team will gain more insight into how the variables interact and what are appropriate changes for improvement. Statistical process control (SPC) is the fundamental tool used for this

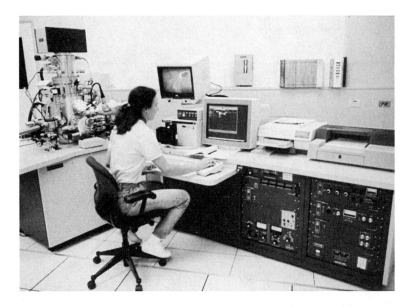

The collection of data requires precise measurement of product variables.
Photo courtesy of Sematech Archives.

analysis. Statistics for experimentation is oriented around design of experiments (DOE), which will not be covered in this book. Statistics for test operations is descriptive, such as how many parts failed a particular test operation.

9.2   STATISTICAL DATA

For manufacturing process control, we collect numerical information, or **data,** about a small number of parts using a **sampling plan,** analyze these data statistically, and then make predictions on how the process is performing. To exploit statistics to its fullest, we first must learn about data.

There are two basic types of data collected in manufacturing:

- *Variables data*
- *Attribute data*

Variables Data

Variables data are characteristics that can be measured. Examples of measurement are thickness, length, temperature, speed, and purity. In each case, a measurement instrument is used to measure the characteristic value. These data are referred to as variables data because the data can vary depending on the measurement results. For example, a wafer is measured at a particular test site for film thickness and results in 8010 angstroms (angstroms is 10^{-10} meters, and is used frequently as a unit for thin film thickness in semiconductor technology). If this measurement is repeated at the same test site on another wafer, and the result is 8014 angstroms, then this thickness measurement is variables data. These measurement data can vary to take on any value.

Process improvement typically involves analyzing process or equipment parameters for optimization. Variables data collected from these parameters provide the most insight for analysis. When statistically analyzed, variables data can tell us how close the desired condition is (in other words, they tell us how good is good). For instance, if a process cuts aluminum rods to be 12 inches long, then we can measure cut rods and obtain variables data to know how close we are. If we measure a rod at 11.9 inches and another rod at 11.5 inches, then we have information about how good the rod lengths are (the 11.9-inch rod *is better* than the 11.5-inch-long rod).

Attribute Data

Attribute data involve assigning some characteristic to a part: tall or short, good or bad, pass or fail. The facts are inherent to the item being measured, and are usually binary. For instance, if we have 100 wafers, and 90 are good and 10 are bad, then 90 percent good and 10 percent bad are attribute data. Test operations in manufacturing frequently use attribute data.

Revisiting our aluminum rod example, attribute data only state that the two cut rods are both short (because both are shorter than 12 inches). There is no information about how much shorter they are, or if one is closer to the desired length. Attribute data provide less information for improvement effort.

9.2.1 Variables versus Attribute Data

To illustrate the difference between variables and attribute data, take an example of walking along a cliff at night while unable to see the edge of the cliff because of darkness. If all you have available is attribute information, then you do not know how close you are to the cliff's edge (you may be walking 10 inches or 10 feet from the cliff's edge, but this information is not available to you). Attribute information only specifies that you have not fallen off the cliff *after each step,* not how close you are to the edge. You only know after each step whether or not you fall off the cliff's edge. You can see how attribute data do not provide information for you to adjust where you are walking until it is too late (you have fallen off the cliff).

On the other hand, variables data provide specific numerical measurements for how far away the cliff's edge is (such as 12 inches, 15 inches, 3 feet). This valuable information is continuously analyzed while walking to predict the likelihood of falling off before the accident occurs, permitting in-process corrections to make sure the cliff never becomes too close. Real-time corrective action based on variables data keeps you from falling off the cliff.

Table 9.1 Variables versus Attribute Data

Parameter	Variables Data (unit)	Attribute Data
Wire Bond Attachment	Pull Strength (force)	Good or Bad
Film Thickness	Thickness (angstroms)	Thick or Thin
Critical Dimension	Width (microns)	Pass or Fail
Gas Flow into Chamber	Flow Rate (torr-liter/second)	High or Low
Particulate Contamination	Quantity (particles/m^2)	High or Low

Because of the binary nature of attribute data (e.g., on/off, good/bad), they have limited usefulness for process control and prediction. For this reason, attribute data will not be covered in this book; however, this is not to say that attribute data should not be collected and analyzed in manufacturing. These data are an important measurement for monitoring conditions such as process yield, equipment performance, or product performance. They can provide a convenient source of process data to support team decisions regarding improvement. The essential aspects of attribute data and different attribute control charts are presented in Appendix 1 at the end of this book.

The information in Table 9.1 summarizes the difference between variables and attribute data for typical manufacturing measurements.

9.3 VARIATION

All processes vary, but none vary the same way. No two natural objects are exactly alike: some humans are shorter and others are taller; some car motors will last longer than others. It is this natural variation that interests us, because if we can reduce variation in a process, then it is more repeatable. If a process is more repeatable, then the equipment works better, and we anticipate producing good parts. A repeatable process is ready for further continual improvement.

The source of all variation in manufacturing entails the four elements discussed in Chapter 1: people, machines, methods, and materials. These elements manifest themselves through their countless variables, all which interact and contribute to variation. A key team goal is to understand these variables, which ones are important for the process being studied, and how they interact.

9.3.1 Common Cause Variation

The natural behavior of any process is based on **common cause variation.** This variation occurs randomly in a predictable manner. These natural sources of random chance are always present in processes. For example, when you drive a car, the slight back-and-forth motion of the steering wheel is natural to the car. This motion is common cause variability.

It is possible to reduce common cause variability, but it does not involve repairing something that is broken. It usually requires understanding the process and equipment to therefore reduce it. We will learn in Chapter 11 how to reduce common cause variation to force the process to continually improve.

9.3.2 Special Cause Variation

The unnatural behavior of a process is **special cause variation.** This is nonrandom variation with an **assignable cause.** Back to the car-driving example, if suddenly the steering wheel started violently shaking, then this would be a special cause. We need to investigate this shaking to determine why it occurred, and we may find that we have a flat tire. To remove the special cause, we fix the flat tire, and then the steering performs normally (the process is now affected only by common cause variation).

Finding the special cause variation usually involves fixing an equipment malfunction or improving an incorrect process parameter. It is similar to finding the root cause to a problem. In manufacturing, correcting special cause variation requires immediate team action to identify and correct the special cause. Once corrected, the process returns to common cause variation with its predictable behavior.

9.3.3 Process Stability

When a process operates only with common cause variation, it is a **stable process.** This process is **in statistical control,** operating with natural variation. Its performance is predictable with statistics, which is necessary for improvement.

If special cause variation is present, the process does not perform in a predictable manner. We typically lack knowledge about the special cause, making it hard to correct. It is now an **unstable process,** or **out of statistical control.** *The only way to change an unstable process into a stable one is to identify and correct the special causes of variation, leaving only common cause variation.*

9.3.4 Natural Process Variation

Every process has its own natural variation. To illustrate, consider three different processes: (1) the toss of a single coin, (2) the toss of two coins, and (3) the toss of three coins.

First, we must define ***probability,*** which is the chance that a certain **outcome** will occur, given all the possible outcomes from a certain action. Mathematically, probability is defined as:

$$\text{Probability of an Outcome} = \frac{\text{Number of Times a Specific Outcome Can Occur}}{\text{Total Possible Outcomes}}$$

First, consider the toss of a single coin. The total possible outcomes from the toss are two, either heads or tails. The probability of any one specific outcome (say, getting a head) is 0.5, or 50 percent, and is calculated as follows:

$$\text{Probability of Heads} = \frac{\text{Number of Times a Head Can Occur}}{\text{Total Possible Outcomes}} = \frac{1}{2} = 0.5$$

It helps to create a table with all possible outcomes and their associated probability to make sure none are overlooked, as shown in Table 9.2.

The sum of probabilities must add up to one (which means that all outcomes are being considered). Because the probability of outcomes for a single coin toss has an equal number of heads and tails, there is a **uniform outcome,** as shown in Figure 9.1.

A **large sample** is required to obtain this uniform outcome. It would be easy to flip a coin twice and have a head each time, which is not uniform. Only with a sufficiently large sample can we attain the confidence that the data reflect the predicted probability of outcomes.

Now let's consider our second process: tossing two coins simultaneously. First, make a table with all possible outcomes, as shown in Table 9.3.

If order is not important (that is, HT is equivalent to TH, which is usually true for a coin toss), then the outcomes are shown in Figure 9.2. This distribution is no longer uniform and is starting to show **central tendency.** Because it is a different process from the first example, the variation is different with three possible outcomes.

Finally, let's consider the process of tossing three coins simultaneously. Table 9.4 shows all possible outcomes. To construct this table, start with HHH, and then move *T* over one space and always fill up with *T* behind it before moving again.

If we again assume that order is not important (e.g., HHT is equivalent to THH), then the probability of outcomes is shown in Figure 9.3. The natural variation of this

Table 9.2 Probability of Outcomes for a Single Coin Toss

Outcomes	H (Heads)	T (Tails)
Probability	0.5	0.5

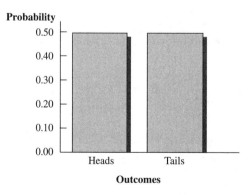

Figure 9.1 Uniform Outcome for a Single Coin Toss

Table 9.3 Probability of Outcomes for a Two-Coin Toss

Outcomes	HH	TH	HT	TT
Probability	0.25	0.25	0.25	0.25

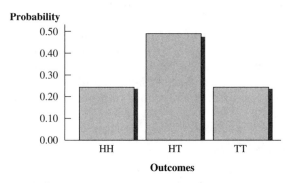

Figure 9.2 Probability of Outcomes for a Two-Coin Toss

Table 9.4 Probability of Outcomes for a Three-Coin Toss

Outcomes	HHH	HHT	HTH	HTT	THH	THT	TTH	TTT
Probability	0.125	0.125	0.125	0.125	0.125	0.125	0.125	0.125

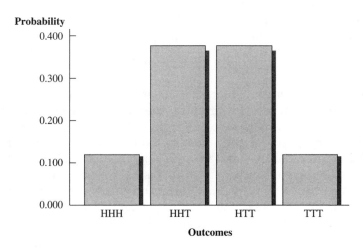

Figure 9.3 Probability of Outcomes for a Three-Coin Toss

process with three coins has four possible outcomes and exhibits more central tendency of the data.

These three different tests of coin tosses are three separate processes, each with its own natural variation. With the single-coin toss, all outcomes are equally likely, meaning heads or tails could occur with any coin flip. As the number of coins flipped increases, so does the chance of getting a mixed outcome (a head and a tail). There is less chance of only getting heads or tails on all three coins simultaneously.

It is typical in nature that different processes vary in different ways. This same variation occurs in manufacturing, with each process having its own natural variation. The goal in manufacturing is to understand this process variation, in order to predict outcomes to ultimately reduce variability with more repeatable results. Successful prediction is a key part of continuous improvement.

9.4 DATA COLLECTION

Statistical analysis requires numerical measurements from variables data collected in the process. These measurement values are called data. Each individual piece of collected data is termed a *data point* within a population. We define how we collect data with a sampling plan, which must clearly state the conditions to ensure the data are collected properly to represent the population.

A population can be either the **total population (N),** which is all potential data that can be measured, or the **sample population (n),** which are the data points that are being measured specifically during the data collection. In other words, the letter *n* represents the sample size.

To illustrate populations, consider the average height of all community college students in the United States. The total population (N) is all community college students. If we take a sample of 200 students and measure their height, then the sample population (n) is the specific 200 students measured to obtain the data.

9.4.1 Manufacturing Data

Manufacturing data are the measurements collected at workstations to provide numerical data about process variables. The data are used for analysis and prediction of future process performance.

Let's consider some manufacturing data. Table 9.5 shows thickness data measured for a film deposited on a wafer during processing. The thickness data are measured in units of ohms per square, which is known as a sheet resistance that correlates the resistance of the metal film to thickness. The thickness data were collected at predetermined test sites (ten per wafer) on six different wafers from the same lot, processed through the same tool, and measured with the same instrument.

Table 9.5 Raw Data for Wafer Film Thickness Measurements (ohms per square)

Wafer Test Site	Wafer 1	Wafer 2	Wafer 3	Wafer 4	Wafer 5	Wafer 6
1	19.8	20.5	19.4	19.7	19.1	20.6
2	20.1	20.0	19.7	19.4	20.0	19.8
3	19.5	20.2	20.2	20.5	19.6	19.9
4	19.7	19.6	19.9	19.8	19.9	20.3
5	19.0	20.0	19.5	20.0	19.3	19.5
6	20.3	20.3	20.2	19.5	19.8	20.0
7	19.9	20.1	20.0	19.3	19.7	19.6
8	19.3	19.6	19.2	19.6	20.2	19.8
9	19.9	20.7	19.4	19.9	19.9	20.1
10	20.4	19.8	20.4	20.1	19.2	19.7

Certain stipulations were placed on how these data were collected: from the same test sites on different wafers from the same lot, processed through the same tool, and measured with the same instrument. Different data collection procedures will change results dramatically. For instance, if we use different measurement instruments, then we have to be concerned about correlation of the data between the two instruments (e.g., they may not be calibrated the same, which is the reason for calibration standards). As another example, if every operator can choose what ten sites to measure on the wafer, then this introduces new variability. If this variability is not of interest (introduced by randomly selecting measurement sites around the wafer), then the operators do not choose, and instead this variable is fixed by defining the measurement sites.

Also notice that for these data, we can analyze the variation in thickness several ways: thickness by all sites on a specific wafer, for an individual test site on all six wafers, or for all test sites combined (which represents the lot). If we compare film thickness at the same test site, we expect less variation, because thickness is always measured at the same location on the different wafers. In essence, it depends on what type of information the team is trying to find from the data. Sometimes it is informative to analyze the data by holding different variables constant just to see the results (check and see if it makes sense based on prior data).

If we vary a variable by measuring different levels (such as varying the amount of time a wafer is in a film deposition tool), and see a change in the measurement data, then this indicates that the variable of time is important for film thickness. There is interaction between these two variables (time and film thickness) that may be important. We state that a **correlation** exists between these two variables. Thus, a longer time in the tool may correlate to a thicker metal film on the wafer. Statistics quantifies this correlation into a numerical relationship.

Production measurement tools measure different product variables.
Photo courtesy of Sematech Archives.

9.4.2 Collecting Good Data

Good data imply that the sample data collected from the sample plan correctly represent the total population. For statistical process control, the objective is to collect good data that represent the inherent variability in the process. Thus, if excessive variation occurs, then a special cause requires team corrective action to identify and correct.

Biased data misrepresent the intended population. Data bias occurs from many sources, such as equipment setup, uncontrolled process conditions, or from inaccurate measurement procedures. Proper data collection technique is important for interpreting statistical results correctly.

As an example of biased data, let's go back to the average student height example at community colleges. If by chance we measure the height of 200 students who are also basketball players, then the data are biased. It is true that these basketball players are also 200 community college students, but the data are collected from only a subset of the total population. This subset does not represent the total population of interest.

In manufacturing, data are usually collected using a sampling plan. A sample plan defines the conditions for collecting data, and will be specified to the operator in the manufacturing procedure. The five requirements for collecting good manufacturing data are:

1. Conduct a random sample.
2. Obtain a large sample.
3. Avoid measurement error.
4. Do not disturb the process during measurement.
5. Use a control group (reference group).

The main technique to avoid data bias is to have a **random sample,** which means that every member of the total population has an equal chance of being selected for measurement in the sample group. Using random sampling techniques, such as selecting parts based on a random location, ensures that the data collected represent the population. Note, however, that for manufacturing process control, personnel are sometimes interested in the repeatability of a certain parameter from run to run. For instance, by always doing measurements at the same location on a wafer, the results are expected to be repeatable (e.g., measuring the same ten locations on a test wafer for controlling film thickness from run to run). In this case, the tool and process variables behave randomly.

A large sample ensures that all members of the population are represented in the sample. Avoid **measurement errors** that lead to incorrect data (incorrectly calibrated gauges, poor instrument repeatability, etc.). Do not disturb the process during measurement, or the changed process will bias the data. And finally, a control group alerts us of any changes to the sample population that occur during the testing.

Data collection requires experience to reduce the chance of getting bad data. The team should seek expert advice for guidance during data collection, including knowledgeable team members or statisticians outside of the team.

9.5 DATA ANALYSIS: HISTOGRAM

The simplest way to analyze raw data is with a visual method known as a **histogram.** A histogram is a plot that categorizes data into intervals, creating a picture of how individual data points are related to one another.

Construct a histogram by grouping the data into categories known as class intervals (or cells), and then plot the data to show the frequency of occurrence for each class. A class interval defines how the data are categorized on the horizontal axis of the histogram plot. Select a class interval width consistent with the level of detail needed to make a decision, given the total spread of the data. For instance, if the data are measured in one-tenth ohm per square (as are our data), then do not choose a class interval of ohms per square. This latter class interval is insensitive to the measured data. In this case, a class interval of one-tenth ohm per square is more appropriate.

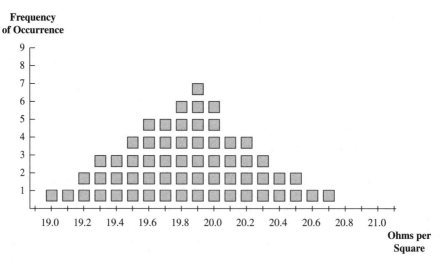

Figure 9.4 Histogram for Wafer Film Thickness Measurements (ohms per square)

A histogram for the thin film measurement data in Table 9.5 is shown in Figure 9.4. We can already draw conclusions about these data (which were not obvious from the raw data). Based on this histogram, if you had to estimate the most likely film thickness measurement, you would probably pick the value with the most measurements: 19.9 ohms per square.

A histogram plot visually highlights the relationship between the different data points. It is not by chance that most manufacturing variables data will have a similar shape as this histogram. The measurement value in the histogram with the most frequent number of measurements is 19.9 ohms per square. In manufacturing, most measured variables are built to a **nominal dimension,** which is where engineering specifications define the target dimension. In the ideal case, the center of the measured values is the same as the nominal dimension, but this is not always true.

There are also **specification limits** for most variables manufacturing data, which define the maximum and minimum acceptable values for the data. The maximum acceptable value is called the **upper spec limit (USL),** and the minimum acceptable data value the **lower spec limit (LSL).**

Assume this wafer film thickness must meet the following specification for thickness:

Nominal Dimension:	20.0 ohms per square
LSL:	19.0 ohms per square
USL:	21.0 ohms per square

These specification limits are shown on the histogram in Figure 9.5.

The data shown in Figure 9.5 have a tendency to group toward the center with data skew toward the lower specification limit. When data are skewed, they lack a

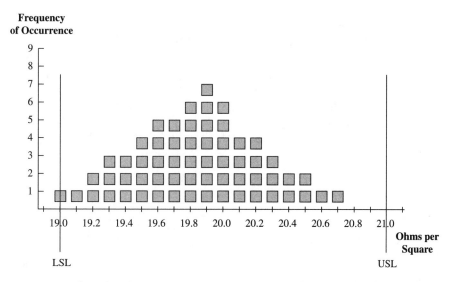

Figure 9.5 Wafer Film Thickness Histogram with Specification Limits (ohms per square)

uniform shape and may not be completely centered, which means the average of the data will probably not equal the nominal specification dimension.

An observation from this histogram is that all of the data are in specification, which is good. However, notice that the lower end of the data is at the edge of the specification limit, while the upper end is farther away from the USL. It is undesirable to be close to any side of the specification limits (USL or LSL), because product produced outside of the specification is defective. The data in this example are not necessarily bad, since no parts are scrapped. But why manufacture a product so close to the specification limits? This is similar to walking close to a cliff's edge at night— why take a chance?

In manufacturing, if the specification limits are far away from the data limits, then there is a **process window.** It is preferable to manufacture with a large process window to permit process variability. A large process window makes it more unlikely to go beyond the specification limits.

Histograms give insight for data interpretation. Different histogram shapes are shown in Figure 9.6. The centered and capable process illustrates a desirable process window.

From a practical standpoint, nearly all variables data for manufacturing processes will have a similar form as these histograms, as long as the sample size is sufficiently large. This occurs because a manufacturing process produces parts to achieve a nominal specification dimension, while trying to stay away from the specification limits (USL and LSL). The goal is for the data to have a repeatable shape in order to reduce process variation so that all parts consistently meet specification. This optimizes the process to achieve repeatable results in high-volume manufacturing.

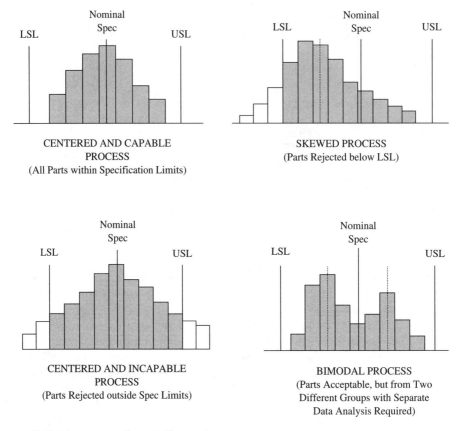

Figure 9.6 Histograms from Different Processes

Plotting data systematically in a histogram yields rich interpretations about the process performance. However, histograms are limited for extensive numerical analysis. For this, we will learn about the normal curve.

9.6 NORMAL CURVE

For numerical statistical analysis, histogram data convert into a continuous curve known as the **normal curve.** The normal curve is also termed the **normal probability distribution,** and represents the predicted relationship of the sample population data. Interpret the terms *curve* or *distribution* as a way to show how measured data are naturally distributed.

Manufacturing data are converted from a histogram to a normal probability distribution by continuously increasing the sample size while reducing the class interval

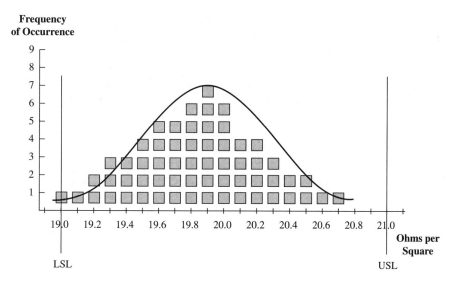

Figure 9.7 Normal Curve for Wafer Film Thickness

size. This process involves considerable mathematics, but can also be understood by simply observing that as the class interval on the horizontal axis is reduced, the width of each interval also becomes smaller. By plotting the center point of each successively smaller class limit, eventually a **bell-shaped curve** is attained, which is the normal probability distribution for the data. Remember, this is only true for data that follow a normal distribution, which is the case for most variables data in manufacturing.

By collecting more data and reducing the class interval size, eventually all the population data fit under the normal curve. In other words, the normal curve represents data collected from the sample population. As shown in Figure 9.7, the normal curve is derived from the sample data. Even though the normal curve is based on sample data, we typically accept that it represents the total population (usually because there are no data to contradict this assumption).

Because variables data are measured data and can theoretically take on all values, the normal curve goes from $-\infty$ to $+\infty$, which explains why the two tails of a normal probability distribution never end. In essence, then, the probability of all the events under the curve is 1.0. Every possible outcome for any particular process being modeled must fit under this curve.

For manufacturing processes, the normal probability distribution accurately describes literally all variables data measurements, because variables manufacturing data exhibit central tendency due to the effort to build to a nominal dimension. The terms *normal curve, normal distribution,* or *bell curve* are interchangeable, and are used to describe the normal probability distribution of the data. The normal curve is also referred to as the Gaussian distribution, named after the famous mathematician Carl Friedrich Gauss.

A normal curve for the wafer thickness measurements is shown in Figure 9.8. This normal curve is from the same data as the histogram in Figure 9.4. Remember that ultimately a normal curve always derives from collected data.

The shape of the normal curve reflects the shape of the data. As the sample size becomes larger and larger, then the normal data more truly reflect a normal curve. If most of the data are near the center, then the curve is narrower. If most of the data are far from the center, then the curve is flatter.

A question that often arises is just how large is an acceptable sample size to accurately represent the true population. The extent of the answer is usually limited in manufacturing by the number of available parts or the amount of time that can be spent on measuring. Nevertheless, the correct statistical sample size can be determined by defining certain sample conditions. If the team needs to know the optimum sample size, consult a statistician. A general rule of thumb (frequently employed in real-life situations) is to have at least thirty data points in the sample.

9.7 CENTRAL TENDENCY

With increasing sample size, manufacturing data tend to centralize around a certain value, termed central tendency. This is an important characteristic of the normal distribution. The most common measurement to describe central tendency for variables data with a normal curve is the **mean** ($\bar{x}$). Another term for the mean is **average.** Interpret the mean as a unique number that best represents the population. If the data are normal, as are most manufacturing variables data, then measurements will tend to centralize around the mean.

To calculate the mean of a group of data, add all individual values from the raw data, and divide by the number of measurements in the sample. This can be expressed mathematically as:

$$Mean\ (x) = \frac{\Sigma\ x_i}{n}$$

where

Σ = the Greek symbol for summation. Data points from 1 to n are summed.

x_i = the value for each individual data point in the sample ($x_1, x_2, \ldots, x_n$).

n = the number of data points in the sample population.

The normal curve and its mean for the wafer film thickness data are shown in Figure 9.8.

Modern calculators typically have statistical functions that will automatically calculate the mean if data are entered properly. For efficiency, students should learn how to use the statistical function on their calculator to perform statistics.

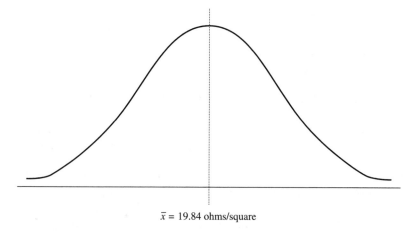

$\bar{x} = 19.84$ ohms/square

Figure 9.8 Normal Curve of Wafer Film Thickness with Mean

There are two other methods for describing central tendency, the median and the mode. The **median** is the midpoint of all data points, and the **mode** is the most frequent data point. Median is used to represent central tendency when no measurement for variability is used (variability is discussed in the next section). An example is real estate. Take the situation when ten houses are for sale, with nine priced at $100,000, while the tenth house costs $10,000,000. The average value overstates dramatically the typical house price (the average is $1,090,000), whereas the median predicts the typical house price as $100,000. The median works better in this case to represent housing prices to a buyer; however, the median and mode are not used frequently in manufacturing for process improvement. These terms therefore will not be discussed further in this book.

This example highlights the use of statistics to correctly represent a population. Any set of statistics can be manipulated to misrepresent reality. Take the situation of the person who drowned in a lake with an average depth of 12 inches. How could this be? The long lake is 11.9 inches deep, except for one small hole that is 10 feet deep. *Misrepresenting reality with statistical data only hinders continual improvement.*

9.8 VARIABILITY

Variability describes how the data are dispersed about the mean. Normally distributed variables data have central tendency about the mean, yet also exhibit dispersion, or spread, about this mean. Variables data rarely have one measured value. If this were the case, flipping a coin fifty times would always yield heads. This outcome is statistically unlikely.

The two primary methods for describing data variability are the range and the standard deviation (sigma).

9.8.1 Range

Range is the difference between the largest and smallest data points of a sample. It is the simplest method to describe the variability of data points about their mean. The formula to calculate range is:

$$\text{Range} = \text{Maximum Data Value} - \text{Minimum Data Value}$$

If all data points have the same value, then the range is zero. There is no variability in the data, which is unlikely in a large sample from a normally distributed population.

To calculate the range for the wafer film thickness data from Table 9.5, find the largest data point and subtract the smallest data point, as follows:

$$\text{Range} = 20.7 \text{ ohms/square} - 19.0 \text{ ohms/square}$$
$$= 1.7 \text{ ohms/square}$$

Interpret the range as the maximum spread about the mean for the different measurements of wafer film thickness. This range of 1.7 ohms per square indicates the maximum distance between data points, but gives no indication where most of the data points lie. To obtain this information, calculate the standard deviation (sigma) and apply certain probability concepts shown in Section 9.9.

9.8.2 Standard Deviation (Sigma)

The most useful term for describing data variability about the mean is **standard deviation,** which describes numerically how the sample data are dispersed about the central point. Standard deviation is represented various ways, including the letter *s,* the symbol σ', or by *SD*.

When referring to the total population, the variability is termed **sigma** (represented as σ). In practice, sigma is commonly used instead of standard deviation for any population, total or sample. By using sigma to describe data, the implication is that the sample population variability equals that of the total population (which is a nice condition, but typically only true with large sample sizes). We will follow this practice and use either sigma or standard deviation to describe variation.

If sigma is a large number, then the data have more spread about the mean. If the data are closely centered on the mean, then sigma is a smaller value. For manufacturing process control, small sigma values are preferred because of less data variability. With a small sigma, the data are more repeatable and thus indicate a more stable process. This relationship is shown in Figure 9.9.

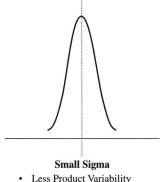

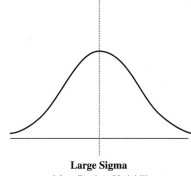

Small Sigma
- Less Product Variability
- More Repeatable Process

Large Sigma
- More Product Variability
- Less Repeatable Process

Figure 9.9 Small versus Large Sigma

Sigma involves more calculation steps than range. The equation is:

$$Sigma\ (\sigma) = \sqrt{\frac{\sum (x_i - \bar{x})^2}{n}}$$

where

$\sum()$ = sum of the values in parentheses.

$(x_i - \bar{x})$ = the difference between each individual data point and the mean of the sample.

n = the sample size (used for a large sample).

This formula is rarely used, because most calculators have a statistical function that calculates sigma with ease. Learn how to enter the statistical function on your calculator, enter the data points, then use the sigma key and the value will be displayed.

The mean and sigma, as shown in Figure 9.10, usually accompany a normal curve. The normal curve is fully defined once the mean and sigma are calculated, which makes sense, because the mean specifies where the center point is, and the sigma describes how the data are dispersed about the mean.

Sigma is specified in terms of the sample size (n) or ($n - 1$), because either one of these may be used in the denominator of the sigma equation. Look on your calculator and you will usually see both σ_n and σ_{n-1} on the keys. For some calculators, the letter s is used in place of σ_{n-1}, meaning this represents the standard deviation. The reason for ($n - 1$) is to compensate for small sample sizes (because it is harder

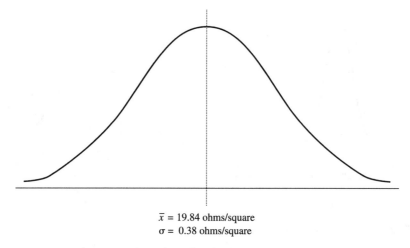

$\bar{x} = 19.84$ ohms/square
$\sigma = 0.38$ ohms/square

Figure 9.10 Normal Curve of Wafer Film Thickness with Mean and Sigma

to determine variability with few data points). For large sample sizes, σ_n and σ_{n-1} will tend to equal each other (there is only 0.1 percent difference between 1000 and 999, but a 10 percent difference between 10 and 9). Any time there is doubt, use $(n-1)$ because this is a more conservative estimate of variability.

Sigma is one of the most important, yet misunderstood, aspects of normal probability. Let's learn why this concept is so critical to understanding variation.

9.9   THE NORMAL CURVE AND SIGMA

The normal curve represents the natural behavior and variation of almost all measured data in manufacturing, because most measured variables are built to a nominal dimension, with upper and lower specification limits. Parts manufactured outside the specification limits are defective. This type of process is best represented by the normal distribution.

The normal curve is completely defined by the mean and sigma of the population. Why? Specifying the mean defines the central point. Sigma describes whether the data are close to the mean (small sigma), or dispersed farther from the mean (large sigma), as previously shown in Figure 9.9.

To understand how sigma represents data variability, remember that 100 percent of all data fit under the normal curve. The probability of all events under the curve is 1.0. Thus, how the area is distributed under the curve is directly related to the probability of occurrence of the data. When the curve hugs the centerline, there is more probability of the data being found close to the mean. When

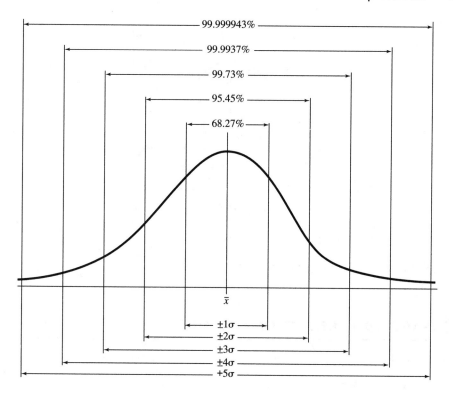

Figure 9.11 Percentages of the Normal Distribution from 1σ through 5σ

the curve is low and spread out, there is less probability of data close to the mean.

The relationship between area under the curve and probability of occurrence for data is used to calculate how many data points are within a certain percentage of the mean. This calculation is done with the normal curve equation, which is not given here.

For a normal distribution, most of the data fall within a few sigma of the mean. As we go additional "sigma" away from the mean, there is progressively less and less data for each sigma. This is seen with a normal curve, because the curve tapers to a low value at its limits. A normal curve with the percentages associated with each sigma is shown in Figure 9.11.

It is useful to know the percentages of a population found within a specified sigma from the mean. Sigma is now used to numerically specify how close the data are to the mean. A small sigma value has data closer to the mean, and with the probability information in Figure 9.11, the percentage of the total population within a sigma value is known.

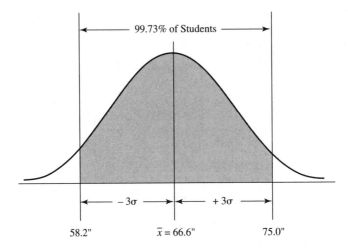

Figure 9.12 $\pm 3\sigma$ of Student Heights

 EXAMPLE 9.1 HEIGHTS OF STUDENTS

We measure the height of thirty community college students and find:

$\bar{x} = 66.6''$ $\sigma = 2.8''$

What is the range of height of 99.7 percent of the students, based on these data?

Solution

Look at the normal curve with percentages in Figure 9.11, and note that their height will be within $\pm 3\sigma$ (this corresponds to 99.7 percent). The range of student heights corresponding to 99.7 percent of the students is calculated as:

$$\bar{x} + 3\sigma = 66.6 + 3(2.8) = 66.6 + 8.4 = 75.0''$$
$$\bar{x} - 3\sigma = 66.6 - 3(2.8) = 66.6 - 8.4 = 58.2''$$

The data predict that 99.7 percent of the students are between a height of 58.2 inches and 75.0 inches. This corresponds to a variability of $\pm 3\sigma$. These results are shown in Figure 9.12.

 EXAMPLE 9.2 PROBABILITY OF FAILURE

You are a maintenance technician, and have ten tools that use a filament as a light source. You want to estimate when to best change filaments. Filaments that fail during a production run affect the product quality and lead to scrap parts. At the same time, filaments are expensive, so you want to avoid changing them early.

You have been collecting the following data on failure rates for filaments:

$$\bar{x} = 670 \text{ hours to failure}$$
$$\sigma = 38 \text{ hours}$$
$$n = 32 \text{ filaments}$$

You know that 3σ corresponds to 99.73 percent probability for a normal distribution. When would you have to change the filaments to have only 0.135 percent fail?

Solution

You recognize that 0.135 percent is one-half of the area outside of $+/-3\sigma$ (found by calculating: $((100 - 99.73) \div 2)$. Sketch a normal curve to show where filaments should be changed to avoid further failures.

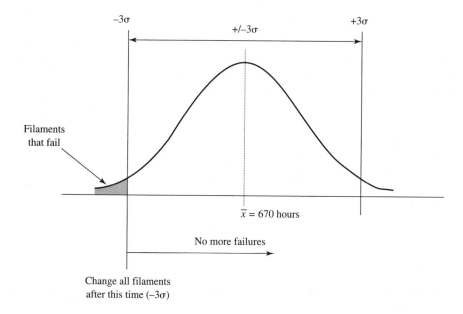

Calculate when the filaments should be changed:

$$\text{Filament change time} = (\bar{x} - 3\sigma = 670 - 3(38)) = 556 \text{ hours}$$

If the filaments in all the tools are changed when they have 556 hours, then the probability of a filament failing during use is 0.135 percent (or 1350 ppm).

The practicality of the sigma calculation is that when a sigma level is specified (such as 4σ, which you should interpret as ±4σ), we automatically know how much of our population is included (99.994 percent for 4σ). Think about this for manufacturing. If a sample population demonstrates that +/–4σ of the data are within the specification limits, then we state that 99.994 percent of our parts meet the specification. Conversely, for this situation, there are 0.006 percent defective parts, or 60 parts per million (ppm) defective, that do not meet specification (derived from the following calculation: 1.0 – 0.99994 = 0.00006 defects, which is equal to a 60 ppm defect rate).

This knowledge is significant for describing a manufacturing process. If someone states they have a 3-sigma process, we know to interpret this as the number of +/– sigma. This corresponds to a percent of the population that meets specification (in this case, 3 sigma corresponds to 99.73 percent, as seen in Figure 9.11).

Sigma can also be interpreted as the number of defective parts outside of its corresponding percentage. Using sigma to describe the number of parts defective (usually in parts per million, or ppm) has become an industry standard for specifying how much of the product meets specification.

Table 9.6 specifies the number of parts defective for a given sigma level. These data come directly from Figure 9.11.

These probabilities are based on the equation for the normal curve, which we have not stated in this book. If the process is normal (as are most manufacturing situations), then for instance, a 6-sigma process produces only 0.002 ppm parts defective (or 2 parts per billion, or ppb). Any assumption that might be used to change this failure rate for a particular sigma level must be clearly stated and justified, and only applies to the process in question.

Some practical advice for calculating the mean and sigma for a population is to ask yourself if the answers make sense. For example, the $\bar{x} \pm 3\sigma$ calculation for the height of students should represent reality. If the calculation found that 99.7 percent of the population was between 24 inches and 96 inches, then we know this is too much variability for the height of people, and there is a problem (e.g., incorrect data, wrong calculation).

Table 9.6 Number of Defects for a Stated Sigma Level

Specification Limit	Parts within Specification	Defective Parts
(+/– σ)	(%)	(ppm outside spec)
1 sigma	68.27%	317,300
2 sigma	95.45%	45,500
3 sigma	99.73%	2,700
4 sigma	99.9937189%	63
5 sigma	99.999943%	0.57
6 sigma	99.9999998%	0.002

You should always be cautious when using a sample population to make predictions about a total population. Qualify predictions based on the soundness of the collected data, plus any particular assumptions used. In the example of student heights, the sample population of fourteen students is used to predict the height of all thirty students in the class (the total population). The data collection techniques must ensure randomness (i.e., all students have an equal probability of being selected), plus recognize that a population of fourteen students is a small sample group.

9.9.1 Probability of an Event

Because of the correlation between the area under the normal curve and the probability of occurrence, the normal curve is used to make predictions about the probability of an event. This event is referred to as y. Do not become confused. This event is called y for convenience because it can be any event associated with the data. Because we do not have the data yet, or know what the event value is, we call it y.

The event y can be to the left or right of the mean. If it occurs on one side of the mean, it is a single-tailed event. A double-tailed event occurs to both the left and right of the mean.

Figure 9.13 shows the area under the curve corresponding to the probability of all the data points greater than the event y. To calculate this probability, it is necessary to know how many sigma y is from the mean. This number of sigma is termed Z. It is also shown in Figure 9.13.

If one knows how many sigma the event y is found from the mean, then based on the normal probability curve, the probability of any occurrence greater than or less than y can be calculated. To find how many sigma the event y is from the mean, use a formula known as the **Z statistic:**

$$Z = \frac{|y - \bar{x}|}{\sigma}$$

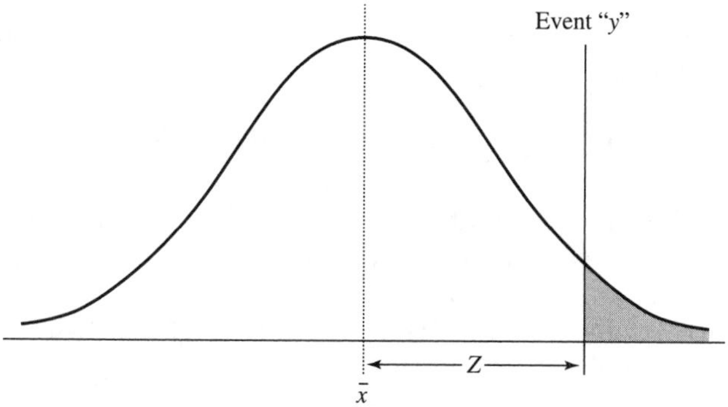

Figure 9.13 Normal Curve with a Single-Tailed Event y

The Z statistic formula is simply a ratio that determines how many sigma the event y is away from the mean. If $Z = 3$, interpret this to mean the event y is 3 sigma away from the mean.

This has practical applications for normally distributed data. For instance, let's reconsider our example of college student heights, where we found the following data:

$$\bar{x} = 66.6''$$

$$\sigma = 2.8''$$

Based on this data, we may want to know what is the probability of having a student in our class taller than 6 feet (72 inches). In this case, y is now equal to 72 inches. We are interested in the probability of all heights *greater than* 72 inches. This probability is the darkened area in the normal curve shown in Figure 9.13.

To find the probability of a student taller than 72 inches, we will find Z (the number of sigma 72 inches is away from the mean), and then use the value of Z in the Standard Normal Probability Table in Appendix 3. This table gives the probability corresponding to a Z value for a single-tailed event.

First, find the Z statistic value:

$$Z = \frac{|72 - 66.6|}{2.8} = 1.93$$

The value 1.93 represents the number of sigma a height of 72 inches is away from the mean, which is based on the sample data we collected for college student heights. Because the formula is a ratio, it does not have units.

Second, relate this Z statistic value to the area under the normal curve that remains to the right of 72 inches (or the event y), as shown in Figure 9.13. Use the Standard Normal Probability Table in Appendix 3 to find this area. The area found translates directly into the probability of the event.

Looking at the Standard Normal Probability Table in Appendix 3, find the number 1.93 ("1.9" in the left hand column and "x.x3" on the top row). Once located, this value corresponds to the probability of students taller than 72 inches. It is converted into percent when multiplied by 100, as:

$$Area_{(Z=1.93)} = 0.0268 \times 100 = 2.68\%$$

The probability of finding a student taller than 72 inches is 2.68 percent. This finding is important because we can now make statements about the general population based on data collected in the sample.

The probability of 2.68 percent is the area to the right of the event y on a normal curve, as shown in Figure 9.13. Note this same calculation can be for some event y that is to the left of the mean or to the right of the mean. If you want to know the probability of both events (for example, a student taller than 72 inches and shorter than 55 inches), then perform the calculations separately and add them together.

A final comment on probabilities is to always keep in mind that any statistical prediction is based on the soundness of data from the sample population. If the sample population data are biased, then the prediction is biased.

 EXAMPLE 9.3   PROBABILITY OF AN EVENT

Using the filament failure rate data in Example 9.2, what is the probability that a filament works up to 800 hours? The data are repeated here for convenience.

$$\bar{x} = 670 \text{ hours} \qquad \sigma = 38 \text{ hours}$$

Solution

The event y is 800 hours. We are interested in the probability that a filament fails after 800 hours (therefore it works up to 800 hours). Use the Z statistic to find the number of sigma the event y is from the mean:

$$Z = \frac{|y - \bar{x}|}{\sigma} = \frac{|800 \text{ hours} - 670 \text{ hours}|}{38 \text{ hours}} = \frac{130 \text{ hours}}{38 \text{ hours}} = 3.42$$

Note that the Z statistic is dimensionless because it is a ratio. Use the Standard Normal Probability Table in Appendix 3 to look up 3.42 and find the area under the curve (which corresponds to the probability of occurrence):

$$\text{Area}_{(Z=3.42)} = 0.00031 = 0.031\%$$

The probability that any filament will last beyond 800 hours is 0.031 percent, which is the same as saying roughly 3 filaments out of 10,000 will last beyond 800 hours. In other words, very few will last this long.

9.10   STATISTICAL VIEW OF THE PROCESS

Statistics is a tool to fundamentally understand how the elements and their variables interact. Once understood, the team can modify process variable settings to achieve optimum goals. Table 9.7 shows how variation, control, and stability are related in manufacturing.

9.10.1 Process Control

If a process has only common cause variation (random or natural), then the process is deemed in statistical control. This process is therefore stable. Outcomes

Table 9.7 Process Variation, Control, and Stability

Type of Process Variation	Control Level	Stability
Common Cause (random)	In Statistical Control	Stable
Special Cause (nonrandom)	Out of Statistical Control	Unstable

from a stable process are predictable. A stable process is efficient, because the manufacturing team can predict efficiency, quality, and production schedules.

If a process has special cause variation (nonrandom or unnatural), then the process is out of statistical control. This process is therefore unstable. The process is inefficient because its output is unpredictable. Special cause variation must be identified and immediately corrected through team action.

New processes are typically unstable. The team goal is to improve any unstable process to stability through the eight-step improvement plan. Once a process is stable (in statistical control with only common cause variation), our intuition might be to stop making changes. However, the work is not done, as we can never stop improving. This would be contrary to continual improvement.

Figure 9.14 depicts a process from a statistics viewpoint.[1] Statistics support the holistic process view, using data to represent the process variables. This view starts with a single data point, grows to a histogram and normal curve as more data are collected, and ultimately represents all process variables through normal distributions. Process instability is created by special cause variation, which inhibits prediction and leads to confusion. As special cause variation is identified and corrected by the team, a process moves toward stability.

The manufacturing challenge is not simply to attain a stable process by working in the process (cyclic effort from the process platform model), but also to move the process toward continual improvement (linear effort). Thus, team complacency toward improvement because a process has attained stability is unacceptable.

To continually improve, a stable process is forced into instability, the team identifies and corrects the new special causes, and makes improvements to restabilize the process. The tool that forces incremental improvement is statistical process control (SPC). We will learn how to use and interpret SPC to continually improve in Chapters 10 and 11.

9.10.2 Process Capability

The other aspect of a process under improvement is **process capability.** Capability depicts how a process varies with respect to the specification limits. A normal curve that represents a capable process is shown in Figure 9.15.

---

[1]See Reference 7 in the bibliography.

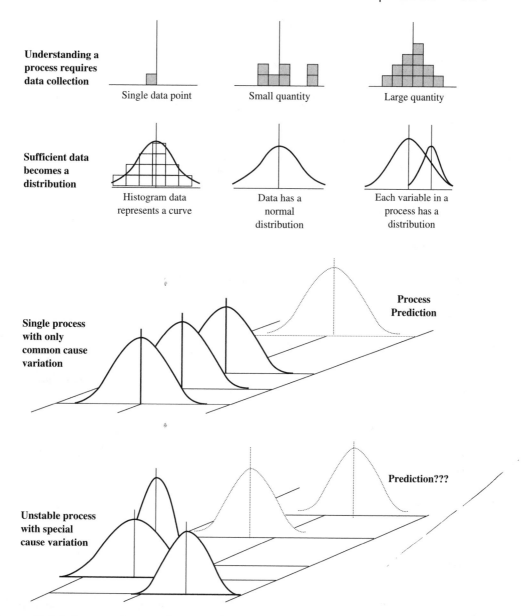

Figure 9.14 Statistical View of a Manufacturing Process

In a **capable process,** all products are produced within the upper and lower specification limits, which is desirable because no defective parts are predicted (remember, the prediction is only as good as the data). However, the goal is not only to have a capable process but also to know *how* capable the process is (how

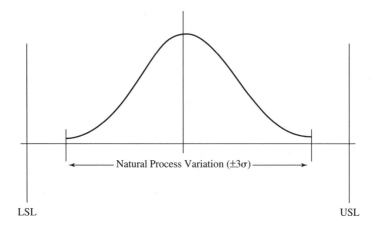

Figure 9.15 Normal Distribution and Process Capability

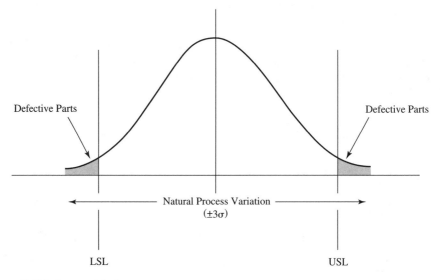

Figure 9.16 Incapable Process

big the process window is). Numerical calculations of capability are discussed in Chapter 12.

A process is not always capable of meeting the specification. An **incapable process** is shown in Figure 9.16. If a process is not capable, then defective parts are built outside the specification limits. The number of defective parts depends on how incapable the process is. For an incapable process, either stop the process to correct the problem or implement short-term solutions to find and rework all defective parts while the long-term solution is implemented. New processes are usually unstable and not capable, requiring intensive team effort to make them capable.

SUMMARY

Statistics is the collection, analysis, and interpretation of numerical data from a population. Manufacturing process control uses statistical data to predict how a process will perform. The two basic types of data are attribute and variables. Variables data are the most important for continual improvement. Data vary by either common (natural) cause or special (unnatural) cause, and a stable process has only common cause variation. Every process has its own natural variation.

Manufacturing data must be properly collected to obtain the correct information and avoid bias. The histogram is a convenient tool for quick visual information about how data are related. Almost all manufacturing variables data for process control are normally distributed, and are represented by calculating the mean and standard deviation (sigma). The mean represents central tendency and sigma represents dispersion about the mean. A large sigma represents more dispersion, whereas a small sigma indicates less dispersion. The normal curve associates a probability of occurrence for any value of sigma from the mean, which can be used to calculate the probability of an event using the Z statistic. A process can be viewed as a statistical relationship between all variables, which permits prediction. This provides the basis for statistical process control and process capability.

IMPORTANT TERMS

Statistics
Population
Data
Sampling plan
Variables data
Attribute data
Common cause variation
Special cause variation
Assignable cause
Stable process
In statistical control
Unstable process
Out of statistical control
Probability
Outcome
Uniform outcome
Large sample
Central tendency
Total population

Sample population
Manufacturing data
Correlation
Good data
Random sample
Measurement error
Histogram
Nominal dimension
Specification limits
Upper spec limit (USL)
Lower spec limit (LSL)
Process window
Normal curve
Normal probability distribution
Bell-shaped curve
Mean
Average
Median
Mode

Variability   Z statistic
Range   Process capability
Standard deviation   Capable process
Sigma   Incapable process

REVIEW QUESTIONS

1. What is statistics, and why is it important?
2. What are the three primary reasons for using statistics in manufacturing?
3. List and explain the two basic types of data. Give two examples of each one.
4. Discuss the different types of information gained from the two types of data.
5. List and explain the two types of variation. State each type of variation in three different ways.
6. What is process stability?
7. When is a process in control?
8. Explain how the probability of an outcome is related to the natural variation of a process.
9. What are manufacturing data and how are they collected?
10. State the five rules for collecting good data.
11. What is a histogram and how is it used to characterize data?
12. What are specification limits in manufacturing, and what are the acronyms?
13. What is a manufacturing process window, and why is it desirable?
14. What is the normal curve?
15. How are manufacturing data converted from a histogram to a normal curve?
16. What is the area under the curve for a normal probability distribution?
17. State four other names for the normal curve.
18. What is central tendency?
19. State the equation for the mean (or average).
20. What is variability?
21. Describe how to calculate the range for a set of data.
22. What is standard deviation? Give four symbols for standard deviation.
23. How is sigma related to standard deviation? What is the symbol for sigma?
24. Draw a normal curve for a small sigma and large sigma, and label each one.
25. What is the most straightforward way to calculate sigma or standard deviation?

26. Explain the difference between σ_n and σ_{n-1}.
27. What two parameters completely define a normal probability distribution?
28. What is the probability of all events under the normal curve?
29. What is the probability of occurrence for $+/-1\sigma$, $+/-2\sigma$, $+/-3\sigma$, $+/-4\sigma$, $+/-5\sigma$, and $+/-6\sigma$?
30. What are the defective parts in ppm associated with the sigma levels stated in the previous problem?
31. Explain how the Z statistic is used to determine the probability of occurrence of an event.
32. Discuss the statistical view of the process in terms of variation, control level, and stability.
33. Explain what a capable process is.
34. What does it mean if a process is incapable?

EXERCISES

Statistical Data

1. Given the following measurements, determine which ones are variables or attribute data.
   A. Wire pull test
   B. Density of a metal-film layer
   C. Number of defective wafers produced at a workstation
   D. Sheet resistance reading
   E. Concentration of dopant material in silicon
   F. Number of good die on a wafer at wafer sort test
   G. Number of tools that meet the availability target

2. You are a maintenance technician in a wafer lab, and have been asked to assess the number of tools that meet the availability targets. As a starting point, you decide to simply list all the tools that meet or exceed their availability, and call these tools acceptable. All other tools that do not meet their availability target are unacceptable.
   A. Explain why these data are attribute data.
   B. Devise a more informative system for assessing how the tools perform with respect to availability by using variables data.

Variation

3. How many total possible outcomes are there when you roll a die? What is the probability of getting a three when the die is rolled? Draw a chart that shows the probability of outcomes for rolling one die. (Note: there are six sides on a die, numbered from one to six.)

4. A bag of marbles contains three blue, five red, and two yellow marbles. If one marble is randomly drawn from the bag, what is the probability of getting a red marble?

5. A deck of cards has fifty-two cards, with four of each type of card (four aces, four kings, etc.). If one card is drawn from a well-shuffled deck, what is the probability that it is a queen?

Data Collection

6. Your team is setting up a sampling plan to measure the sheet resistance at different sites on a monitor wafer from each lot. Based on experience, you know that the film layer is typically thicker in the middle of the wafer than at the edges. The amount of time to measure is important, because the data verify whether the lot of parts is acceptable to proceed in production.

    Consider the following approaches to setting up a sample plan. What are the positive and negative aspects of each approach, and will they both work for this sampling plan? Which one is the best sampling method for process control? Why?

    A. Measure as many points possible in the time allotted, and let each operator randomly select the measurement sites.
    B. Define certain measurement sites across the wafer, ensuring that the center and edges are represented. These same sites are measured on each monitor wafer.

7. Review the following data collection plans for a manufacturing process and discuss what could potentially give biased data:

    A. Two different operators use two separate measuring tools to collect data.
    B. You are assessing the performance of parts from a tool. There are three different suppliers of material used in the tool, but you are not distinguishing between them for your measurements.
    C. For inspection purposes, you always select the top wafer in a cassette of twenty-five wafers.
    D. Five hundred wafers are processed in a 12-hour shift, but you only measure one wafer due to shortage of time.
    E. A sampling plan is established to measure film thickness on wafers from two different tools. During the run, you decide to make some adjustments to a tool.

Histogram

8. Each student in the class shall make at least thirty measurements from a sample population of his or her choice from outside the classroom. The student determines what is measured, but it must be variables data. Be careful to collect good data. Record all the raw data, specifying:

    • the variable measured (ensure that it is variables data), and population size
    • how the measurements were done, and units of measure

A. Make a histogram plot of the data. Give the plot a title and label both axes. Place the correct units on the axes.
B. Do the data exhibit central tendency?

9. You are controlling the concentration of a dopant material implanted in a silicon wafer by measuring the resistance of the silicon after implantation with a four-point probe. You obtain the following values (in ohms) on a sample of ten wafers, with five measurement sites per wafer (one measurement in the center, and four measurements at the outer edge):

Data for Resistance Measurements to Control Dopant Concentration (ohms)

Wafer #	Site 1	Site 2	Site 3	Site 4	Site 5
1	704	701	700	706	699
2	697	700	700	697	697
3	699	700	698	697	701
4	698	702	701	699	700
5	706	703	699	698	699
6	704	698	700	700	704
7	701	700	700	700	701
8	703	701	703	696	698
9	702	697	697	695	695
10	703	699	699	700	694

Make a histogram plot of these data. Label the histogram and the axes.

10. For the histogram data in Exercise 9, the specification limits are:

USL: 710 ohms
LSL: 690 ohms
Nominal: 700 ohms

A. Describe the data relative to the specification. Are the data centered in the specification, or skewed toward one side?

B. Are all parts in specification?

C. Would you estimate the average of the total sample close to the nominal dimension?

Central Tendency and Variability

11. What is the average and the sigma (σ) for the histogram data in the previous exercise? Use the statistical function of your calculator to find these values.

12. Let's assume that we have a time constraint and can only do one measurement per wafer in Exercise 9. We can only measure the first measurement in each row for each wafer (Site 1), for a total of ten measurements.

A. Recalculate the average and sigma using only these ten values from Site 1 (use σ_{n-1} because of the small sample).

B. Calculate the percentage difference between the average of the total sample (all fifty sites) and this new sample. Note: *% difference* = {(*total* − *new*) / *total*} × 100.

13. We have more time available for measurement, and are able to do three per wafer in Exercise 9. We obtain the first three measurements in each row, which are Sites 1, 2, and 3, for a total of 15 measurements.

A. Recalculate the average and sigma using only these fifteen values (again, use σ_{n-1}).

B. Calculate the percentage difference between the average for the total sample of fifty measurements and this new sample.

C. Has the percentage difference improved from Exercise 12(b)? Explain why it should have improved.

14. Use the statistical function on your calculator to calculate the mean, sigma, and range for the following data that represent the heights of students in a college class (inches):

64″, 67″, 65″, 71″, 68″, 62″, 68″, 70″, 68″, 66″, 69″, 60″, 68″, 66″, 67″, 67″

Total number of students in the class: N = 30

Sample size: n = 16

Normal Curve and Variability

15. For the variables data collected by each student in Exercise 8, calculate the mean, sigma, and range.

16. How many defective parts (in percent and in ppm) do the following processes produce?

3 sigma = _____ % = _____ ppm defective

4 sigma = _____ % = _____ ppm defective

5 sigma = _____ % = _____ ppm defective

6 sigma = _____ % = _____ ppm defective

17. For the heights of community college students measured in Exercise 14, at what heights will 95.5 percent (2 sigma) of the students be found?

18. For the heights of community college students measured in Exercise 14, what is the probability of finding a student taller than 6 feet? (Hint: Use the Z statistic.)

19. For Example 9.2, a business review of the data determines that we must achieve a single-tailed, +/– 4-sigma process. This means that only 0.0032 percent of the filaments will fail in use.

   A. Calculate at what time the filaments should be changed to achieve this failure rate (use the data supplied in the problem).

   B. If we continually shorten the time to change filaments to achieve a lower failure rate, then we throw away more and more good filaments. With this logic, we could put all filaments in for zero seconds, therefore none fail, and we have a zero failure rate. What do you recommend as a way to overcome this problem?

20. Recalculate Exercise 19, but now assume that improvement activity has reduced the sigma for the time to failure by 75 percent to 9.5 hours. The mean time to failure that is stated in Example 9.2 stays the same. Is the time to change filaments sooner or later than in Exercise 19? Why?

21. Since the beginning of record keeping, rainfall for the month of August in Austin, Texas, is shown below. Based on these data, what is the probability of getting 10 inches of rain or more during the month of August in Austin?

$$\bar{x} = 5.5 \text{ inches}$$

$$\sigma = 2.8 \text{ inches}$$

22. Given the same rainfall data as in Exercise 21, what is the combined probability of getting less than 1 inch or greater than 10 inches of rain in Austin, Texas, during the month of August?

23. Based on the wafer thin film thickness data shown in Figure 9.10, calculate the probability of finding:

   A. A film thickness greater than 21.1 ohms/square

   B. A film thickness less than 18.9 ohms/square

   C. A film thickness outside of the following specifications:

   USL = 21.1 ohms/square

   LSL = 18.9 ohms/square

   D. Draw a sketch of a normal curve, and label the mean, sigma, and specification limits.

   E. Identify the reject areas of the curve by darkening in the appropriate areas and label the amount of parts rejected (in percent).

10

✳ STATISTICAL PROCESS CONTROL CHARTS

Statistical process control (SPC) is an analytical tool for understanding fundamental process performance. With SPC, teams develop the wisdom to properly identify problems, find potential root causes, verify the correctness of changes, and force the process toward continual improvement.

With basic statistical knowledge, we will now learn how to build an SPC chart using variables data. This chart is termed the x-bar and range chart, or $\bar{x}$ & R chart, and is valid for analyzing any variables measurement data from the equipment, process, or product.

_____ **OBJECTIVES** _____

After studying the material in this chapter, you should be able to:

1. List and discuss the five benefits for using SPC charts.
2. Explain the most common type of control charts used for variables data.
3. Explain the concept of an SPC chart and its advantage for presenting data.
4. Explain how special and common cause variation are indicated on a control chart, and how they are related to control limits and out-of-control situations.
5. Given raw subgroup data, all formulas and tables, and a blank SPC chart, construct a complete SPC chart.

10.1 INTRODUCTION TO SPC CHARTS

SPC charts guide team interpretation of process variables in a manufacturing process. Because of its ability to separate variation into common and special cause, the control chart is a powerful tool to statistically control process performance. With this ability to control the process, manufacturing personnel can consciously guide it toward improvement.

The five benefits from using control charts in manufacturing are:

1. ***Data collection focal point.*** SPC charts provide a comprehensive system for collecting, analyzing, and presenting data for process performance.

2. ***Out-of-control action.*** Control charts alert the team about special cause variation that requires immediate corrective action, and guide the team with visual interpretations of the root cause. This supports the cyclic effort in the eight-step improvement plan.

3. ***Demonstrate process stability.*** Control charts are visual indicators that verify the process is stable (in statistical control), and therefore producing predictable output.

4. ***Drive continual improvement.*** Control charts drive team improvement of the process by reducing control limits to tighter levels. This forces improvement action on previously acceptable variation, moving the process linearly toward continual improvement.

5. ***Assess process capability.*** Control charts identify process control limits, and assist in assessing the process capability to build product that conforms to specification limits.

The most common type of control charts in manufacturing is the variables data control chart for average and range, $\bar{x}$ and R, and the variables control chart for average and standard deviation, $\bar{x}$ and S. We will learn how to build an **x-bar and range** ($\bar{x}$ & R) **chart** in this chapter, and use this chart as a basis to understand and interpret control charts in Chapter 11. The **x-bar and standard deviation** ($\bar{x}$ & S) **chart** supplies essentially the same information as a chart using the range to measure variation, but will detect changes in the data dispersion better. SPC software programs in manufacturing typically use $\bar{x}$ and S because of the simplicity to calculate the standard deviation with a computer. Always use standard deviation for a control chart if the subgroup size is greater than ten. Different types of control charts and their relevancy to manufacturing are summarized in Appendix 1.

Control charts are also used for attribute data, such as the number of defective parts at an inspection operation. There are several types, with the most common known as the p-chart, to control the proportion of parts that do not conform to requirements (defective parts). Attribute control charts are also reviewed in Appendix 1.

Use attribute control charts cautiously, as they can give the false impression that the process is in control, even though defects are present in the process. A process is never in control (statistical or otherwise) if defects are produced.

10.1.1 Concept of an SPC Chart

The conceptual layout of an $\bar{x}$ & R variables control chart for averages and range (known simply as an $\bar{x}$ & R chart) is illustrated in Figure 10.1. For an in-control

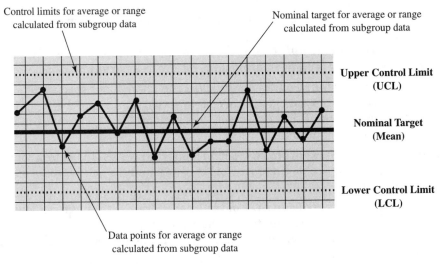

Figure 10.1 Conceptual Layout of an SPC Chart

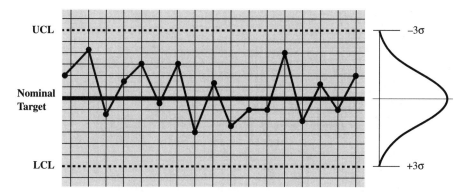

Figure 10.2 Control Chart with Normal Curve

process, data are randomly located with central tendency on both sides of the **nominal target,** and within the **upper control limits (UCL)** and **lower control limits (LCL).** For the same reason that a single coin flip is sometimes heads and sometimes tails, probability predicts certain outcomes depending on the process. Randomly distributed data points exhibit no long runs of data on either side of the nominal target, and no obvious trends in the data (in the same manner as not getting twenty heads in a row from subsequent coin flips, followed by twenty tails in a row).

An SPC chart is a visual representation of statistical data that are normally distributed. Figure 10.2 illustrates how an in-control SPC chart represents the normal curve in a user-friendly manner. The ability to **visually interpret** process variation is a major advantage of control charts. A quick look at Figure 10.2 confirms that the process has no data points outside the control limits, no data point runs, and is therefore in control.

Control limits are not specification limits, and should never be confused as such. SPC limits are calculated based on the data the team collect from the process variable. We will learn to calculate control limits shortly, and will learn in Chapter 12 how control limits interact with specification limits to predict process capability.

10.2 CONTROL CHARTS AND VARIATION

The $\bar{x}$ & R control chart in Figure 10.3 demonstrates how SPC efficiently separates between common and special cause variation. Common cause variation is associated with data points that:

1. plot within the calculated control limits; and
2. randomly plot with central tendency about the nominal target.

Points that follow these patterns represent in-control data. A visual inspection of the control chart in Figure 10.3 reveals nonrandom data on one side of the nominal

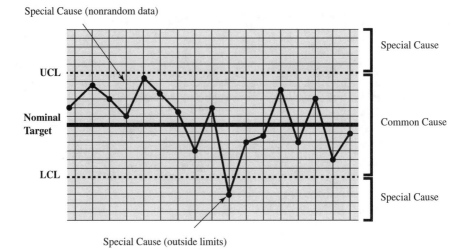

Figure 10.3 Overview of Common and Special Cause Variation

target and one data point outside of the LCL. This process is out of control. **Out-of-control data** derive from special cause variation, requiring immediate corrective action to understand why the situation occurred.

To understand how control limits work, recall our knowledge about the normal distribution. We predict a certain percentage of the outcomes will fall within a defined sigma level, with central tendency around the mean. For instance, at a $\pm 3\sigma$ level, 99.73 percent of the data points are within the sigma limits with central tendency around the nominal target. SPC equations include this probability information to establish control limits, which visually show where data will plot if they are affected by only common cause variation. If the data do not plot as expected, then the process is unstable with special cause variation.

A variables control chart analyzes process data in terms of the average (central tendency) and range (dispersion), and provides insight into how both vary. Therefore $\bar{x}$ & R control charts typically have two separate charts: an average and a range. The $\bar{x}$ (average) chart is situated on top, and the R (range) chart is on bottom. Each chart has its own unique upper and lower control limits that we will soon learn how to calculate.

Control chart data are collected, measured, and analyzed in **subgroups.** The process variables measured in the subgroups are defined by the team and are used to calculate the average, $\bar{x}$, and the range, R. Interpret subgroups as small, random snapshots of the process variable. These subgroup snapshots are the building blocks for establishing process control limits and measuring process performance.

Upper control limits (UCL) and lower control limits (LCL) indicate the maximum permissible variation of subgroup data, and are marked on the charts as dashed lines. The nominal target (mean) represents the central tendency of the subgroup data, and is marked as a solid line. These two parameters, control limits and

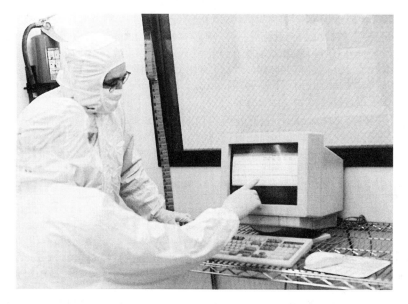

Two operators inspect data relative to control limits on an SPC chart.
Photo courtesy of Ion Implant Services, Inc.

the nominal target, function together to highlight out-of-control data, which is logical because we know that normal data are entirely defined by the mean and sigma, which correspond to the nominal target and control limits, respectively.

10.3 DEVELOPING A VARIABLES SPC CHART

Modern SPC charts are constructed on computer systems, with automated data collection and computers performing the calculations. The final chart with the plotted data and control limits is displayed on a computer screen at the operation where the work is done. Team members interpret the SPC chart, and take corrective action when necessary.

For educational purposes, we will construct an SPC chart manually to understand why SPC data are displayed in their unique format. The team should have already performed steps 1 and 2 of the eight-step improvement plan, and identified the key process variables for SPC analysis.

The five steps for creating an $\bar{x}$ & R control chart are:

1. **Define the subgroup measurements.** SPC data are collected in subgroups for the measured variable. Define subgroups for size and frequency.
2. **Collect and record subgroup data.** This is step 3 of the eight-step plan. SPC data are collected and recorded on the $\bar{x}$ & R chart.

3. **Calculate and plot subgroup data points.** Calculate and plot the average and range for each subgroup as the data are entered.

4. **Calculate and plot nominal targets.** Calculate and plot nominal targets for the average of the subgroup average ($\bar{\bar{x}}$) and the average of the range ($\bar{R}$). Perform this calculation when sufficient subgroup data are available (typically twenty-five sets of subgroup data).

5. **Calculate and plot control limits.** Calculate and plot the UCL and LCL for the $\bar{x}$ (average) and R (range) charts. Perform this calculation when sufficient subgroup data are available (typically twenty-five sets of subgroup data).

Step 1: Define the Subgroup Measurements

Information acquired during the initial analysis stage of the eight-step improvement plan is used to identify the key variables affecting the process. Review each key variable to assess how it will be measured. *Be careful—do not select noncritical variables for SPC analysis.* This leads to waste if unimportant information is collected. Ensure that the data are properly collected, or else uncontrolled variation is introduced into the natural process.

SPC chart data are defined in subgroups, usually with four or five data points in each subgroup. A smaller subgroup size is permissible, but try to have at least three data points. Each subgroup is analyzed as a complete set of data, and will produce a point on the $\bar{x}$ (average) chart and the R (range) chart.

Define subgroups so that the opportunities for variation between each data point in the subgroup are small. For instance, a subgroup may be consecutively produced parts off of a single tool. Because the parts are produced under similar conditions, the variation within a subgroup represents the common cause variability over a short time. If special cause variation occurs, the control chart highlights this by going out of control.

Choose the **subgroup measurement frequency** to detect all process changes over time. Give thought to the nature of the problem to be detected, such as changes between tool setups, shift patterns, relief operators, warm-up trends, material lots, or other factors. Define the sample frequency based on these types of considerations. For instance, if you are concerned about incoming material variability and receive a new material lot every shift, then it would not make sense to measure only once per week. You probably want to measure a sample from each lot.

The frequency between subgroup measurements is typically short for an unstable process. As the process stabilizes, the time interval between subgroup measurements should increase, because less data are needed to verify process control. Reducing unnecessary SPC measurement time increases process efficiency. For example, SPC measurements for an unstable process may be done multiple times per shift. Once special causes have been eliminated, subgroup data collection could be reduced to once per shift.

Relevant information regarding the nature of the measurement procedure is recorded on the $\bar{x}$ & R chart (time of day, specific tool, units, operator, etc.), as shown in Figure 10.4. Do not overlook this step, because it contains valuable information for the corrective action stage later in the improvement effort.

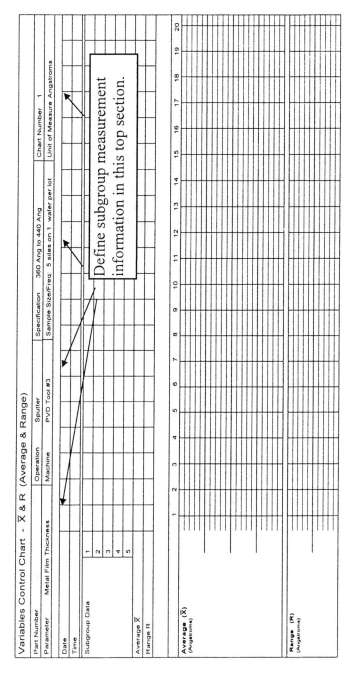

Variables Control Chart - X̄ & R (Average & Range)

| Part Number | | Operation | Sputter | Specification | 360 Ang to 440 Ang | Chart Number | 1 |
| Parameter | Metal Film Thickness | Machine | PVD Tool #3 | Sample Size/Freq: 5 sites on 1 wafer per lot | | Unit of Measure | Angstroms |

Define subgroup measurement information in this top section.

Figure 10.4 Step 1: Define the Subgroup Measurements on an x̄ & R Chart

215

Step 2: Collect and Record Subgroup Data

The team collects subgroup data using the guidelines previously discussed in Chapter 9. Record all relevant events in a process log, such as tool changes and new raw material lots. This information becomes critical during subsequent problem analysis.

Collecting SPC data is time consuming, so avoid measuring unimportant process variables. *The cost of collecting insignificant SPC data is one of the main reasons why firms often ignore SPC.*

Record the collected subgroup data in the appropriate part of the SPC chart, as shown in Figure 10.5. Note that in practice, each time a set of subgroup data is recorded (five data points in this example), the $\bar{x}$ and R statistical calculations are done and the point is plotted. The statistical data in each subgroup serve two purposes: first, to calculate the individual points on the average and range charts, and second, to calculate the control limits.

Step 3: Calculate and Plot Subgroup Data Points

Once subgroup data are entered on the control chart, the $\bar{x}$ (average) and R (range) values are calculated for each subgroup. Perform subgroup calculations as soon as the data are recorded on the chart.

To calculate the **subgroup average** ($\bar{x}$), sum the measurements within a subgroup, and divide by the number of data points in that subgroup. Record this number on the chart under the subgroup. The equation for the subgroup average ($\bar{x}$) is:

$$\bar{x} = \frac{x_1 + x_2 + \cdots + x_n}{n}$$

where

$\bar{x}$ = the average value of subgroup measurements.

$x_1 + x_2 + \cdots + x_n$ = the sum of measurements within one subgroup, from 1 to n subgroup data points.

n = the number of measurements in a subgroup.

Next, calculate the **subgroup range** (R), which is the difference between the maximum and minimum subgroup value, and record this value on the chart. The formula for the subgroup range is:

$$R = x_{maximum} - x_{minimum}$$

where

$x_{maximum}$ = the largest subgroup measurement value.

$x_{minimum}$ = the smallest subgroup measurement value.

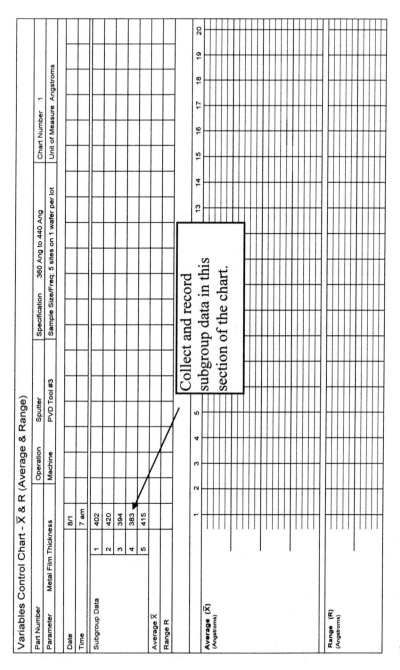

Figure 10.5 Step 2: Collect and Record Subgroup Data on an $\bar{x}$ & R Chart

217

Sample calculations for $\bar{x}$ and R are shown highlighted in the boxes of Figure 10.6. Record the subgroup data for $\bar{x}$ (average) and R (range) on the chart in the subgroup section (see Figure 10.6). Before plotting values on the $\bar{x}$ (average) and R (range) charts, define a suitable scale on the left side of the plot. This scale should be large enough to encompass all expected values for $\bar{x}$ and R. This may be difficult to anticipate at first, so the scale can be modified later after calculating several subgroups.

Plot the value for $\bar{x}$ and R on the corresponding charts. Each subgroup has a corresponding point on their respective $\bar{x}$ (average) and R (range) charts. The points on the two charts align vertically with the subgroup column. As points are plotted, connect them with a continuous line. These lines give a quick visual representation for how the subgroup data statistically vary on the average and range charts.

Initially, the subgroup data for average and range plot on their respective charts with no control limits. This process is acceptable, because of the limited numerical information about the process to calculate correct control limits. The nominal target and control limits are calculated only after obtaining sufficient data (there is no firm rule, but usually after approximately twenty-five separate subgroups have been measured).

Step 4: Calculate and Plot Nominal Targets

Calculate the nominal targets for $\bar{x}$ (average) and R (range). Nominal targets represent the central tendency for both the average and range SPC data. They are also referred to on a control chart as the mean.

The two nominal targets are

$\bar{\bar{x}}$: The mean of the averages for all subgroups used in the calculation.

$\bar{R}$: The mean of the ranges for all subgroups used in the calculation.

We will refer to $\bar{\bar{x}}$ as the nominal target for the average chart. In practice, it has several names: **X-double bar,** or simply mean (or also *mean of the averages,* or *mean of the means,* or *grand mean*). Do not be confused by this terminology. It is simply the average of all the subgroup averages that were already calculated. It represents the best estimate of the nominal target where all subgroup averages should ideally measure.

$\bar{R}$ is the nominal target of the ranges. It is also known as **R-bar** or *range mean.* It is the target value for all ranges calculated from the different subgroups, and represents the central tendency for all subgroup ranges. (Range measures dispersion, but the subgroup ranges have an average dispersion.)

To calculate $\bar{\bar{x}}$ (the average nominal target), use the following formula:

$$\bar{\bar{x}} = \frac{\bar{x}_1 + \bar{x}_2 + \cdots + \bar{x}_k}{k}$$

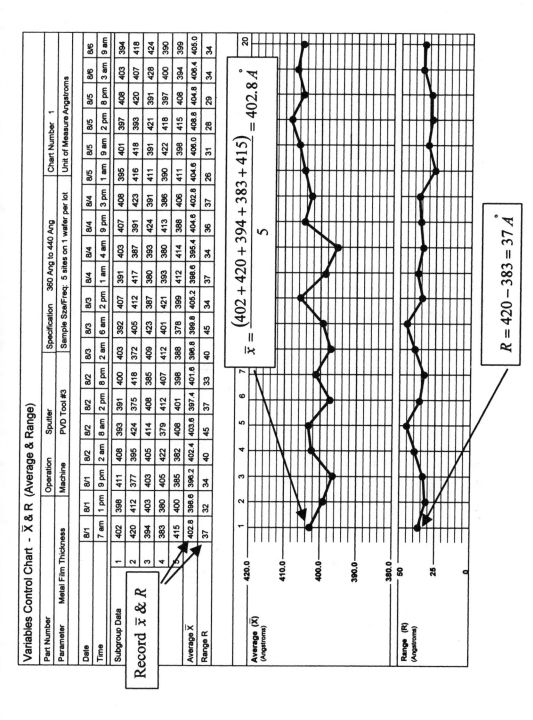

Figure 10.6 Step 3: Calculate and Plot Subgroup Data Points

where

$\bar{\bar{x}}$ = the mean of all subgroup means ($\bar{x}$). Referred to as the nominal target for the average ($\bar{x}$) chart.

$\bar{x}_1$ = the mean of subgroups 1, 2, 3, and so forth, for each subgroup.

k = the number of subgroups chosen to calculate the nominal targets.

The formula to calculate $\bar{R}$ (the range nominal target) is:

$$\bar{R} = \frac{(R_1 + R_2 + \cdots + R_k)}{k}$$

where

$\bar{R}$ = the mean of all the ranges, referred to as the nominal target.

R_1 = the range of subgroup 1, 2, 3, and so forth, for each subgroup.

k = the number of subgroups used to calculate the nominal targets.

For the wafer film thickness data shown in Figure 10.7, $\bar{\bar{x}}$ and $\bar{R}$ are calculated as:

$$\bar{\bar{x}} = \frac{(402.8 + 398.6 + 396.2 + \cdots + 405.0)}{20} = 402.1 \text{ Å}$$

$$\bar{R} = \frac{(37 + 32 + 34 + 40 + \cdots + 34 + 34)}{20} = 35.2 \text{ Å}$$

Once $\bar{\bar{x}}$ and $\bar{R}$ are calculated, they are drawn as solid lines on their respective charts: $\bar{\bar{x}}$ on the average chart, and $\bar{R}$ on the range chart, as shown in Figure 10.7. Label each of them on the left side of their respective line.

The nominal targets ($\bar{\bar{x}}$ and $\bar{R}$) represent the best estimate of central tendency for the average and range based on the subgroup data used in the calculation. In an ideal case, the nominal targets are equal to the nominal specification value. This fact is rarely true, however, leaving it as a team action to work toward shifting the nominal target of the chart toward the specification nominal value over time.

Nominal targets (and control limits) are calculated based on having all process variation represented in the subgroup data, because this represents the actual process condition. The team members judge when they believe this is true. The most straightforward way to determine this is to inspect the control chart and note when the plotted values on the average and range charts do not show erratic behavior.

If significant process variation is omitted when calculating the nominal targets, then the process quickly goes out of control, and is not a desirable situation because

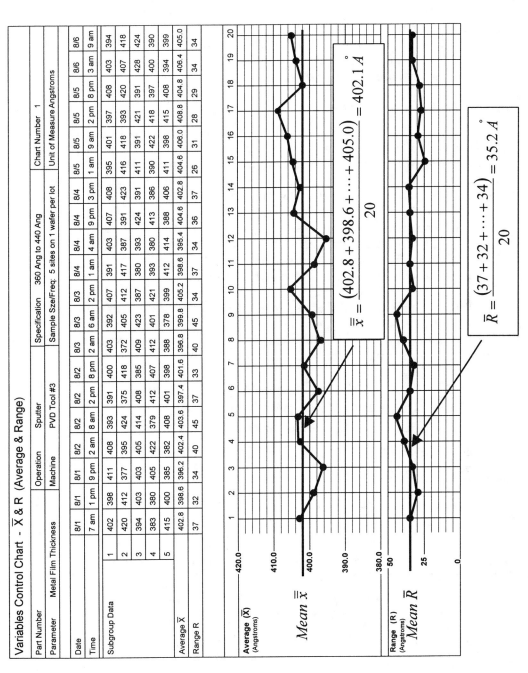

Figure 10.7 Step 4: Calculate and Plot Nominal Targets ($\bar{\bar{x}}$ and $\bar{R}$)

the manufacturing line is continually disrupted for predictable problems. Eventually the SPC charts will be discredited and ignored. A general rule of thumb is to calculate the nominal targets (and control limits) after approximately twenty-five subgroup data sets have been measured.

Step 5: Calculate and Plot Control Limits

Control limits, with the nominal targets, signify the limit of expected statistical variation for the subgroup average and range of the measured variable. Control limits are calculated from multiple subgroup data, and encompass all variation present in the process at the time of their calculation. Therefore, control limits represent the process conditions for the period that their calculation is based.

When viewed in this manner, control limits coupled with the nominal targets are reference points for how a process is expected to statistically perform. Interpreting limits and targets as reference points simply means that they function as a process platform, from which all future process performance is judged until the team decides to change the control chart limits and targets to improve the process.

Control limits and nominal targets are based on process data from multiple subgroups. As with the nominal targets, ideally twenty-five sets of subgroup data are used for calculating control limits. To calculate the upper and lower control limits for the average and range, use the following formulas:

$$UCL_R = D_4\overline{R} \quad \text{(upper control limit for range)}$$
$$LCL_R = D_3\overline{R} \quad \text{(lower control limit for range)}$$
$$UCL_{\overline{X}} = \overline{\overline{x}} + A_2\overline{R} \quad \text{(upper control limit for average)}$$
$$LCL_{\overline{X}} = \overline{\overline{x}} - A_2\overline{R} \quad \text{(lower control limit for average)}$$

where

$D_4, D_3,$ and A_2 = constant values from Table 10.1.

Table 10.1 D_4, D_3, and A_2 Constants for Control Limit Calculations ($\overline{x}$ and R)

	"n" (number of measurements in a subgroup)								
	2	**3**	**4**	**5**	**6**	**7**	**8**	**9**	**10**
D_4	3.268	2.574	2.282	2.114	2.004	1.924	1.864	1.816	1.777
D_3	0	0	0	0	0	0.076	0.136	0.184	0.223
A_2	1.880	1.023	0.729	0.577	0.483	0.419	0.373	0.337	0.308

$$\bar{\bar{x}} = \frac{\bar{x}_1 + \bar{x}_2 + \cdots + \bar{x}_k}{k} \text{ (from step 4)}$$

$$\overline{R} = \frac{(R_1 + R_2 + \cdots + R_k)}{k} \text{(from step 4)}$$

The constant values for calculating the control limit equations are given in Table 10.1. These constants reflect the variability and sample size of each subgroup. The constant values depend on the size of the subgroup, n.

The quantity n from Table 10.1 is the number of measurements in a subgroup. Do not confuse this with k, which is the number of subgroups used to calculate the nominal targets for each chart. The values in Table 10.1 are actually derived from the equation for the normal curve. This table is used in manual SPC calculations as a convenience. If a software version of SPC is used, then all the calculations are done internally on the computer.

The control limit calculations for the wafer film thickness example in Figure 10.8 are:

$$UCL_R = D_4\overline{R} = 2.114 \times 35.2 = 74.4 \text{ Å}$$
$$LCL_R = D_3\overline{R} = 0 \times 35.2 = 0 \text{ Å}$$
$$UCL_{\overline{X}} = \bar{\bar{x}} + A_2\overline{R} = 402.1 + (0.577 \times 35.2) = 422.4 \text{ Å}$$
$$LCL_{\overline{X}} = \bar{\bar{x}} - A_2\overline{R} = 402.1 - (0.577 \times 35.2) = 381.8 \text{ Å}$$

Once calculated, control limit values are drawn as dashed lines on the $\bar{x}$ (average) and R (range) charts, and labeled appropriately (UCL or LCL), as shown in Figure 10.8.

Control limits represent boundaries that statistically predict the behavior of the subgroup data when only common cause variation is present, which is the same manner that a nominal target predicts the behavior for central tendency. Control limits give a visual limit for how far a process is expected to statistically vary based on the subgroup data. If the process varies outside the control limits, then special cause is present. Immediate corrective action is required to identify the reason for the special cause and correct it.

Control limits are calculated only from subgroup data. Specification limits (USL and LSL) are not used to calculate the control limits, which illustrates why control limits represent process conditions during the data collection time period, and are a reflection of process performance. Control limits do not represent the specification limits.

SPC is a visual history of a manufacturing variable over time. The control limits do not change unless the team makes a decision to change them, and are a reference point by which all future performance of this variable is measured. This is why SPC charts function as a process platform, placing a "stake in the ground" to serve as a beacon to the team on the path toward continual improvement.

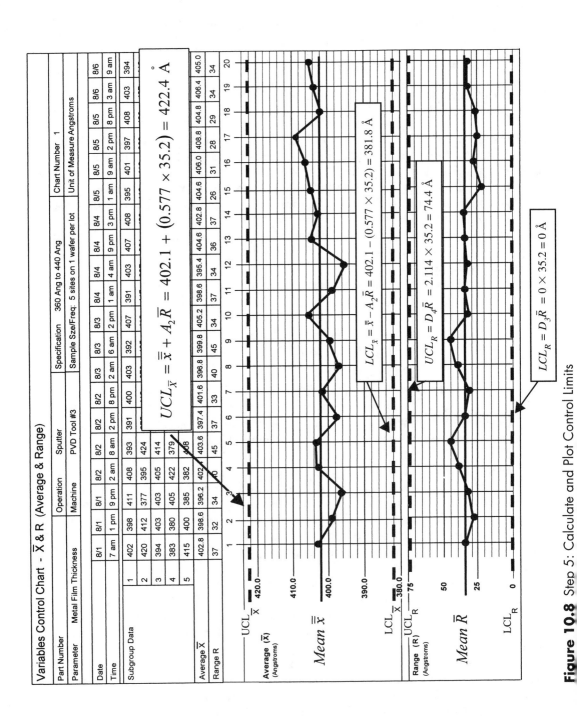

Figure 10.8 Step 5: Calculate and Plot Control Limits

SUMMARY

The five reasons for using SPC control charts in manufacturing are data collection, out-of-control action, process stability, continual improvement, and process capability. Data on a control chart plot randomly with central tendency around the nominal target and do not exceed the control limits. Subgroup data points that do not follow this criteria are out of control with special cause variation present, which gives a visual interpretation to process control. The five steps to manually constructing an SPC chart are to define the subgroup, collect and record the subgroup data, calculate and plot the subgroup data, calculate and plot nominal targets, and finally, calculate and plot the control limits. Measure critical variables for SPC to avoid collecting unnecessary data.

IMPORTANT TERMS

Data collection focal point
Out-of-control action
Process stability
Continual improvement
Process capability
x-bar and range ($\bar{x}$ & R) chart
x-bar and sigma ($\bar{x}$ & S) chart
Nominal target
Upper control limits (UCL)
Lower control limits (LCL)

Visually interpret
Out-of-control data
Subgroups
Subgroup measurement frequency
Subgroup average
Subgroup range
X-double bar
R-bar
Control limits

REVIEW QUESTIONS

1. State and discuss the five benefits for using SPC charts in manufacturing.
2. What is the most common type of control chart in manufacturing?
3. Explain the concept of an SPC chart.
4. Explain how in-control data are located on an SPC chart.
5. Explain how the control limits and nominal target function to highlight out-of-control situations on an SPC chart.
6. State the five steps for creating a variables SPC chart.
7. What is a subgroup, and how do you define subgroups for an SPC chart?
8. Why is subgroup frequency important?
9. Why is it important to record information about the subgroup on the control chart?
10. Why do subgroup data initially plot on a control chart with no limits?

11. What is the rule of thumb for the number of subgroups needed to calculate control limits?
12. What is the nominal target on an SPC chart?
13. What are four different names for the nominal target of the average chart?
14. Explain how much process variation should be included in calculating nominal targets and control limits.
15. Explain the difference between the variable n and k for SPC chart calculations.

EXERCISES

Building SPC Charts

Note: Recall that SPC charts are filled out as soon as the individual subgroup data are measured. For practicality, the following exercises give you multiple sets of subgroup data at once.

1. You are a new employee and working at an operation that is using an SPC chart. While looking over the chart, you notice that the control limits on the chart are the same value as the specification limits. You ask another operator how the limits were calculated, and are told that the engineer makes the charts and no one ever questioned the calculations.

    When you see the engineer, you ask about the control limits. The engineer has never had a course in SPC, and always assumed that upper and lower control limits take on the same value as specification limits.

    Explain the following concepts to the engineer, and how they apply to an SPC chart:

    - Common cause variation
    - Special cause variation
    - Nominal targets
    - Control limits
    - Difference between control limits and specification limits

2. The following subgroup data were collected over the past 2 weeks for a furnace temperature. A team member explains the data were collected once per shift, with

Data for Furnace Temperature (°C)

Subgroup 1	Subgroup 2	Subgroup 3	Subgroup 4	Subgroup 5	Subgroup 6
1200.0°C	1198.9°C	1201.0°C	1200.1°C	1199.8°C	1201.0°C
1200.2	1199.7	1199.8	1201.1	1200.6	1200.4
1200.4	1198.9	1199.8	1201.1	1200.6	1200.4
1199.5	1198.7	1199.9	1200.6	1199.3	1200.2
1201.0	1200.3	1199.9	1200.1	1201.0	1199.8

five measurements per subgroup. The specification limit is 1200 +/– 1.5°C. However, there is some confusion about what to do with the data.

A. You are asked to show your team how to construct an $\bar{x}$ & R variables control chart. Perform all necessary calculations, plot the subgroup data, and draw lines representing the nominal targets and control limits for both the average and range charts. Properly label the chart. Use the six sets of subgroup data provided.

B. What recommendations can you make to the team with regard to the number of subgroups and calculating the control limits? Are there sufficient subgroups for calculating realistic control limits?

3. You work in thin films in a wafer fab (applying thin aluminum layers on the wafer surface with a physical vapor deposition tool). The process requires a wafer as a test monitor included for each run. The aluminum film thickness on the test monitor is measured at five locations and represents the thickness on all wafers in the lot. The measurement data are input into a log to see if any data are out of specification. To better control the process, your team decides to use the data to construct an $\bar{x}$ & R SPC chart. Since you learned about SPC at school, your team members ask you to build the first chart.

Data for Wafer Aluminum Film Thickness, PVD Tool 1 (angstroms)

Date/Time	Test Site 1	Test Site 2	Test Site 3	Test Site 4	Test Site 5
3/11 8:30 A.M.	5,930	5,970	6,060	6,040	5,840
3/11 11:00 A.M.	6,010	5,950	6,040	5,880	5,960
3/11 3:00 P.M.	6,040	6,030	5,940	5,990	6,010
3/11 8:30 P.M.	5,840	6,000	6,050	5,930	6,060
3/12 1:00 A.M.	6,020	5,880	5,920	6,040	6,010
3/12 6:30 A.M.	5,940	6,020	5,890	6,000	6,090
3/12 10:00 A.M.	6,040	5,950	6,100	6,010	5,930
3/12 4:40 P.M.	6,100	5,960	6,050	6,080	5,990
3/13 3:00 A.M.	6,020	6,080	5,940	5,890	6,090
3/13 8:00 A.M.	5,940	5,890	5,840	6,040	6,110
3/13 1:00 P.M.	6,020	6,050	6,050	6,000	5,940
3/13 7:30 P.M.	5,860	5,980	6,020	6,040	5,940
3/14 1:30 A.M.	6,050	6,000	5,940	5,990	6,000
3/14 5:00 A.M.	5,940	6,080	6,050	5,860	5,950
3/14 1:30 P.M.	6,020	6,050	5,940	5,950	6,030
3/14 6:00 P.M.	5,940	5,990	6,020	6,040	6,000
3/15 2:00 A.M.	6,020	6,050	6,010	5,940	5,990
3/15 5:00 A.M.	5,890	5,950	6,010	5,930	5,990
3/15 2:30 P.M.	5,990	5,970	6,040	6,010	5,980
3/15 6:00 P.M.	6,040	6,080	5,980	6,070	6,020

To demonstrate how an SPC chart works to your colleagues, use the thickness data to build a variables $\bar{x}$ & R chart, filling in all data points on the chart and using lines to show the nominal target and control limits for the average and range (label these lines). Put a scale on the chart with units. The specification limits are 6000 +/– 250 angstroms.

4. Construct a variables $\bar{x}$ & R SPC chart using the following data. Fill in all appropriate chart information. Put a scale on the chart and identify the units. Plot the subgroup data. Show all limits and nominal targets as a line with their proper label. The specification is 21 +/– 2.5 ohms per square.

Data for Sheet Resistance for Metal Film Layer, PVD Tool 5   (ohms per square)

Date & Time	Test Site 1	Test Site 2	Test Site 3	Test Site 4
8/3 8:00 A.M.	20.2	21.3	21.1	20.8
8/3 2:30 P.M.	21.6	21.0	20.7	21.5
8/4 3:00 A.M.	21.9	22.5	20.6	22.0
8/4 2:00 P.M.	19.3	20.4	21.2	19.8
8/5 5:00 A.M.	19.9	20.9	20.3	21.3
8/5 1:00 P.M.	21.4	20.7	22.3	22.0
8/6 1:30 A.M.	22.4	23.1	21.4	20.7
8/6 3:00 P.M.	21.0	21.9	22.7	20.6
8/7 4:30 A.M.	19.8	20.6	19.4	21.0
8/7 10:30 P.M.	22.5	19.4	21.3	20.8
8/8 8:00 A.M.	21.3	20.4	22.8	23.2
8/8 5:30 P.M.	19.8	21.3	20.4	19.1
8/9 3:00 A.M.	22.8	23.2	21.0	20.4
8/9 4:30 P.M.	19.0	23.3	21.8	20.1
8/10 8:30 A.M.	22.5	21.4	22.8	21.9
8/10 5:00 P.M.	21.4	20.1	22.7	21.5
8/11 2:30 A.M.	22.3	20.4	21.5	22.7
8/11 3:00 P.M.	19.8	21.3	19.1	20.7
8/12 4:00 A.M.	19.1	22.4	21.3	19.8
8/12 3:30 P.M.	21.5	19.6	20.7	22.4

11

INTERPRETING SPC CHARTS FOR TEAM ACTION

People interpret processes regularly. If your body runs a fever, you look for the cause and treat it. If a bicycle chain keeps slipping off the gear, you check the chain tension. The ability to analyze processes to change and improve is one of the most significant aspects of being a human. We can apply this human skill to manufacturing for dramatic improvement.

The primary way to gather information for process analysis in manufacturing is statistical process control (SPC). Interpreting SPC charts to control a process is an art. When mastered, SPC simplifies the complex interaction between the process elements, clarifying the role of each variable. SPC knowledge gives the team a basis for clear decisions and focused action.

OBJECTIVES

After studying the material in this chapter, you should be able to:

1. Explain why SPC is used with respect to control.
2. State and interpret the three SPC visual rules for assessing if a process is in control, and describe what is required if a process is out of control.
3. Be capable of describing the various data trends on SPC charts, and give possible reasons for their occurrence.
4. Explain how an out-of-control situation could be desirable.
5. Describe the condition for when a production line should be shut down based on SPC chart information.
6. Explain the concepts of shifting mean and increased range.
7. Describe the action necessary to correct special cause variation.
8. Explain how common cause variation is necessary for improvement, and how SPC chart information is used to force a process to improve continually.

11.1 WHY SPC?

Why is SPC used in a process? SPC is used because it guides team actions on improving the process with information about its control. The use of SPC raises the following questions about the process and team actions:

- Is the process in control or out of control?
- If the process is in control, do we stop because the work is done, or do we make additional changes to improve it further?
- If the process is out of control, what action needs to be taken?

How these questions are answered determines whether SPC is used effectively. We will learn the correct answers as we acquire SPC interpretation skills in this chapter.

11.2 CONCEPT OF CONTROL

Process control is the ability to modify a process output to a desired performance. In electromechanical control systems, this is achieved with feedback control, such as a furnace controller that maintains a temperature set point. A temperature sensor monitors the temperature in the furnace (the output), provides feedback with a

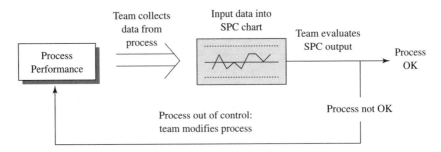

Figure 11.1 SPC Feedback Control System

voltage signal to a controller, which then modifies the process by turning on and off the heater to maintain the desired set point.

For statistical process control, the team monitors the process output by visually inspecting the SPC charts. People then provide feedback through team corrective action to identify and correct special causes to maintain a stable process. *The essential element for* **SPC feedback control** *is team corrective action to modify a process for an out-of-control situation.* The SPC feedback control system is shown in Figure 11.1.

SPC charts demonstrate control through visual indicators that show whether the process is performing as statistically expected. If necessary, the team initiates corrective action to modify the process to the desired performance. The three basic indicators of control on an SPC chart are:

Control Limits: Data plot within the control limits (+/–3 sigma).

Nominal Target: Data are centered about the nominal target (central tendency).

Random Pattern: Data have random trends expected for a normal distribution (dispersion about the mean).

These control indicators are based on the expected outcome for normal probability data, as we learned in Chapter 9. Because nearly all manufacturing process control data are normal, they are stable and in control when these three control indicators are met.

Because the control chart is based on data collected from the process, *the control chart and the process are one and the same.* A poorly performing process has out-of-control SPC charts, whereas a stable process demonstrates stability on the control chart.

11.2.1 Everyday Process Control

Let's consider driving a car on a freeway at regular speed as an example of process control that we do every day. If you are a passenger, you expect the driver to keep the car inside his or her lane. It can move to the left and the right, but the driver should maintain the car toward the center of his or her lane. This is central tendency with random motion on either side of the nominal target (center of lane). The process is in control.

When the driver lets the car drift to one side of the lane, it is unacceptable. The car does not exceed the lane markings (or control limits), but it is not acceptable for the car to drift. At this point, the process is out of control, and corrective action is required. Possibly the car's front end is out of alignment, or the driver is daydreaming, both which may cause unnatural variation (a drifting car). It is important to determine the special cause so that corrective action can be taken, and the car is returned to the center of the lane.

If the car drifts too far in one direction, it goes beyond the lane markings and makes a loud noise by hitting lane reflectors. The driver and passengers are immediately alerted of a problem. Lane markings with reflectors are essentially control limits, designed to highlight when the car is out of control. The driver takes immediate corrective action and returns back to the center of the lane. The passenger looks at the driver to check for drowsiness or other problems (trying to identify a special cause).

This car-driving example illustrates the use of the basic concepts of SPC on a daily basis. The primary difference between this example and the SPC used in manufacturing is that SPC requires numerical data instead of visual observation. These numerical data statistically predict how the process should perform to remain in control.

11.3 PROCESS CONTROL

SPC charts are used to control a process. If a process is in statistical control, then the process is stable with predictable performance. If a process is out of control, then it is unstable with unpredictable performance. An unstable process requires team action with cause-and-effect analysis to identify and correct the special cause that is making it unstable. Corrective action leads to equipment and process improvement.

SPC charts highlight out-of-control situations, but never correct problems. Corrective action is an individual and team responsibility. *If a person does not act to correct the problem highlighted on a control chart, then SPC adds no value and is therefore waste.* There are three **SPC visual rules**[1] for assessing whether a process is out of control with SPC control charts. These rules are shown in Table 11.1.

The key to proper use of these rules is *if the process is out of control, then the team takes action to identify the special cause and correct it.* The amount of action must be appropriate to return the process to an in-control situation. An individual may actually do the corrective action, but the team reviews the action and gives appropriate support. These rules apply to both the average and range charts (one can be out of control and not the other).

The visual rules help to interpret SPC charts for out-of-control situations, yet are not a rigid set of rules. Knowledge for interpreting control charts derives from

---

[1]The actual number of visual rules can vary, but we use three as a practical approach.

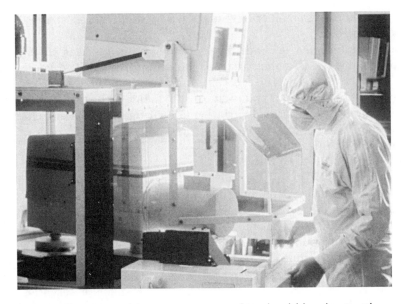

Corrective action to correct problems is necessary, but should be done with error-proofing to eliminate recurring problems.
Photo courtesy of Sematech Archives.

Table 11.1 SPC Visual Rules for Out-of-Control Situations

1. Point outside of control limits:	Point outside of upper or lower control limits on either the average or range chart
2. Run of points:	Run of points (usually seven or more) on either side of the nominal target
3. Nonrandom patterns:	Any obvious nonrandom pattern on the average or range chart that indicates trends

working in the manufacturing process to attain a balance between rules and human insight. With experience, team members understand their process and learn how to appropriately interpret out-of-control data on the chart. The essential ingredient is individual commitment to improve the manufacturing process through team action.

Rule 1: Point Outside of Control Limits

Rule 1 states that a process is out of control if one or more points plot beyond either the upper or lower control limits. This applies to either the average or range chart.

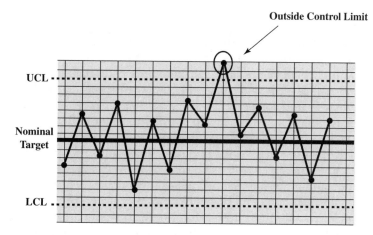

Figure 11.2 Data Point Outside Control Limit

It requires immediate action to identify and correct the special cause that created the out-of-control situation. Figure 11.2 shows a chart with an out-of-control data point outside the control limit. There is still a trend toward central tendency of the data around the nominal target.

If data plot outside the control limits, the operator highlights the out-of-control data on the chart to identify it as special cause variation requiring investigation. The highlight could be a circle around the out-of-control point or a color indicator on a computer screen (SPC computer software does this automatically).

If the source of the problem is understood, then it is corrected and noted in the workstation problem log to inform other team members. It should be discussed with team members at shift changeover or during a team meeting to understand why it occurred, the corrective action, and how error-proofing can be implemented.

If the source of the special cause is not immediately understood, then additional technical support is needed, which may involve notifying other team members such as the line technician, the supervisor, or the engineer. Team action is needed for technical breadth, whereas tools such as brainstorming and cause-and-effect analysis are used to identify and correct the unnatural process variation.

Be aware that sometimes SPC data are out of control for reasons unrelated to the equipment and process, such as:

- The control limit or subgroup data point has been miscalculated.
- There is an error in the measurement procedure.
- The point has been plotted incorrectly.

The team investigation for an out-of-control situation should be within the context of the eight-step improvement plan. Start with obvious solutions and proceed from that point.

Rule 2: Run of Points

Rule 2 states that if data make a run of points on one side of the nominal target of the average or range chart, then the process is out of control. Normally, data randomly plot on both sides of the nominal target while exhibiting central tendency. If this does not occur, there is special cause affecting the process variable that requires team action to identify and correct.

Consider the height of a sample group of college students. The students have an average height, which is represented by the nominal target on an SPC chart. Some students are shorter than the average and others are taller. We would never expect a random selection of students to always be on one side of the nominal target (mean).

This expected randomness of SPC data is shown on the left side of Figure 11.3, with data points plotted above and below the mean. It is acceptable for a few points to fall on one side of the nominal target, but not for too many (typically up to seven points in a row on one side is acceptable, but be sensitive to trends).

On the right side of this control chart, the data plot only above the nominal target. The process is out of control for Rule 2 because the points make a run on only one side of the nominal target. Something has changed in the process. If these data represent the height of college students, then we know the average student height has shifted, and we need to find out why (e.g., the sample population has changed and only student basketball players are selected for measurement).

To return the process in Figure 11.3 to an in-control situation, the team must analyze the process to find the root cause for the shift. Once this special cause is identified and corrected, the process returns to the in-control situation (similar to the left side of the chart).

It is appropriate to question whether returning to the previous condition is good team action. The data in Figure 11.3 shifted to an out-of-control situation because of a run of points above the mean. But, upon closer review, the process variation actually improved after the shift occurred (there is less dispersion among the data points).

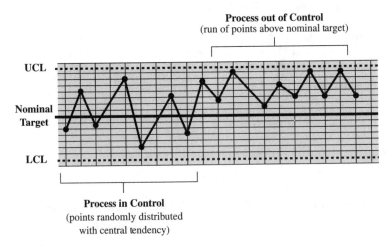

Figure 11.3 Out-of-Control Run of Points

An operator inspects a pattern on an SPC chart.
Photo courtesy of Ion Implant Services, Inc.

The ideal situation would be to return the shifted mean to the nominal target, but retain the reduced variation. This requires personnel to first understand why the shift occurred, and second, what caused the reduced variation. This type of thinking is how SPC can be used to not only control, but also improve, a process.

This analysis also illustrates an important point: *An out-of-control process is not necessarily bad.* In fact, an out-of-control process may actually be moving in a good direction (such as an out-of-control chart for fewer contamination particles in a tool). The important point is to understand *why* an out-of-control situation occurs. If we do not take action to understand why, then it may very well revert back to the worse situation.

Rule 3: Nonrandom Patterns

Rule 3 states that control charts with patterns or trends are out of control. A pattern is a repeatable shape, and a trend usually goes in one direction. Hereafter, both are referred to as a pattern. Patterns reveal nonrandomness and are out of control, even if the data do not exceed the control limits or exhibit a run of points.

Because SPC is a visual tool, patterns are important to visually interpret the process. Patterns demonstrate unexpected process performance by highlighting special cause variation. Learning to interpret patterns can help the team determine the root cause of the unnatural variation. Figure 11.4 illustrates different control chart patterns that apply to either the average or range plot. In addition, Appendix 2 describes ten common control chart patterns and their possible causes.

Team action depends on the shape of the pattern and how we interpret the special cause. Consider pattern 3 in Figure 11.4, which is a shift in level. This is an

1. Cycles

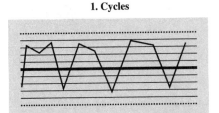

Causes: Rotation of operators, temperature variations, and different parts suppliers

2. Freaks

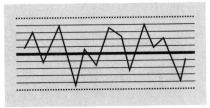

Causes: Occasional point outside of limits with no reason to be found (statistical chance)

3. Shift in Level

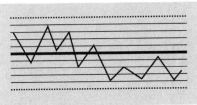

Causes: Gradual introduction of new materials, change in maintenance program

4. Instability

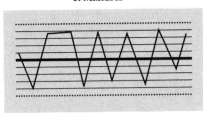

Causes: Overadjustment of machine, improper setup, difference in test sets or gauges, lack of discipline

5. Mixtures

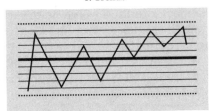

Causes: Combination of 2 different patterns on same chart — 1 high & other low; separate to analyze

6. Trends

Causes: Tool wear, operator fatigue, change in chemical composition, change in temperature

Figure 11.4 Sample Out-of-Control Patterns with Possible Corrective Actions

out-of-control situation according to the third visual rule for nonrandom pattern. We must look for some abrupt change in the process that caused the shift in level, such as a change in material or a new tool setup.

Moving the process back to the original condition is good, but there is more information available to the team. Looking at pattern 3, note the process was shifted slightly higher than the nominal even before the major downward shift occurred. If the reason for the downward shift could be understood, then the team can possibly use this knowledge to actually center the process on the nominal target. *Process knowledge guides the team to improve beyond the initial condition.*

✳ **CASE STUDY** ✳
Control Chart Patterns

Because of the complexity of a particular process in a wafer fab, operators are required to take a tool out of production after a designated period and certify that critical tool parameters were in control. The tool reconditioning involves running test wafers in the tool, measuring certain variables on the test wafers, and then plotting the results on SPC $\bar{x}$ & R control charts to verify the process is in control. If acceptable, the operator continues with production. If out of control, then engineering is notified to correct the problem and approve the tool for production. Because of the impact to production, this tool recertification is often done as a team effort, but sometimes it is done alone (depending on how many operators are available).

An experienced operator noticed during a tool setup that another operator was ignoring bad data by not plotting the facts on the SPC chart. The operator did correctly reevaluate the tool with new test wafers, but *only good data were plotted* on the control chart.

The experienced operator realized the importance of the control chart patterns to assess subtle shifts in tool performance. Omitting data meant that operators on different shifts were not aware of important tool trends over time.

The experienced operator discussed this with the other operator, who agreed it was not good practice, but was reluctant to stop because it would mean calling the engineer to review the out-of-control data. (This delays bringing the tool back on line.) The process engineer was notified, and after a discussion between the operators and engineer, everyone agreed that the correct procedure of logging all data would be followed.

Points to consider

1. What was the fundamental reason for not plotting the bad data on the control chart?

2. Would it have made any difference if the operator had the skills to evaluate and release an out-of-control tool back to production without engineering?

3. Was there an effective team effort to resolve this problem? Does more need to be done?

Avoid overinterpreting data patterns, because even random data can sometimes give the illusion of nonrandomness (special cause). With experience and hands-on knowledge of the equipment, team members learn the subtle relationship between control chart patterns and a potential special cause. Having experience helps to quickly find the root cause of problems, which is important for manufacturing.

Given this close relationship between experience, knowledge, and data interpretation, the importance of long-term commitment and training in manufacturing is evident. During the capacity production period from the 1950s through the 1970s, manufacturing expertise was not important. Manufacturing was viewed as an abstract entity that anyone could do. In the 1990s, many firms still place minimal importance on developing long-term manufacturing expertise, to the detriment of improvement.

11.4   AVERAGE VERSUS RANGE

The two separate charts on an $\bar{x}$ & R variables control chart are the average and the range charts. These reflect the two parameters that define normal data: mean and sigma. The average chart represents central tendency (mean), whereas the range chart represents variation (sigma). If you think about the concepts of central tendency and variation, you appreciate that they are two opposing ideas. Central tendency reflects data measuring toward the center point, whereas variation reflects dispersion of data away from the center point. For this reason, we always want to know the control situation of both the average and the range of the data. The average and range can each independently go out of control or be stable. How? Let's consider a shifting mean.

11.4.1   Shifting Mean

If the mean of a measured variable changes all at once, then the average plot on the SPC chart will reflect the change and the chart will likely go out of control. This is termed a **shifting mean.** If the population average changes slowly over time, then this would be a trend in the mean. While the average data are changing and even going out of control, there can still be a stable range.

Let's look at a common example: cutting wood boards with a table saw. The variables datum is length, measured in inches. A positioning fixture is adjusted on the table saw to cut boards 24 inches long (as shown in Figure 1.2 of Chapter 1). There is some minor variation about this length (every board is not exactly the same length), but overall the length is fairly close. If you were to measure board lengths and plot the data on an SPC chart, the nominal target for the average becomes 24 inches. Both the average plot and range plot are in control.

After cutting numerous boards, you measure a group and find the length has increased by 1 inch. You investigate, and find that the positioning stop on the table saw was moved by a coworker. The mean length of boards has shifted by 1 inch. The variability about the mean is the same, because *the same sources of variability exist,* as shown in Figure 11.5.

If board length measurement data were plotted on an SPC chart, the average chart would go out of control when the length shifted to 25 inches due to a run of

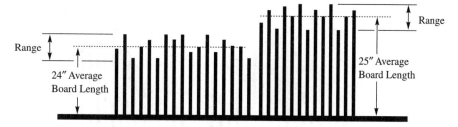

Figure 11.5 Shifting Mean for Boards Cut on a Table Saw (With Stable Range)

points above the mean (the average board length was 24 inches, and now is 25 inches). The range chart remains in control.

A shifting mean is a common occurrence in manufacturing. The shift stems from various sources, such as a different batch of material or using a new fixture to set up a tool. It generally derives from an abrupt change in the process. If a mean shift occurs, then action is required on the process to evaluate why the shift occurs. It may be that the shift is desirable (i.e., it is toward the nominal specification value), but team members must understand it to keep the process from unpredictably changing again.

11.4.2 Increased Range

Now let's consider another scenario: **increased range.** The average of a process can remain stable (all points are in control on the average chart), while the variation of the data about the range nominal target increases until the range is out of control. A process will have less repeatability with an increase in range, which is undesirable.

Let's revisit our table saw example. Assume the table saw positioning stop is set to cut an average board length of 24 inches. As we cut boards, they are neatly cut at the 24-inch length with minimal variation (the process is centered with a small sigma). However, after cutting for some time, we notice the board lengths are different. The lengths are not uniform, with some boards noticeably shorter than other boards. We investigate, and find that the saw blade shaft developed a severe vibration (possibly from a bad bearing). The board lengths now vary substantially, as shown in Figure 11.6. The average length is still 24 inches, but the variation about this mean has increased significantly. Based on these data, the SPC range chart will be out of control for the increased variation in board length, while the average chart remains in control.

Adjusting the positioning stop will not correct the out-of-control range problem. We have to address the root cause, which is a problem related to the saw blade that has a severe blade wobble.

This example illustrates that to find the correct special cause, we want to know whether it is the average or range that is out of control. This information guides us to the correct root cause for taking corrective action.

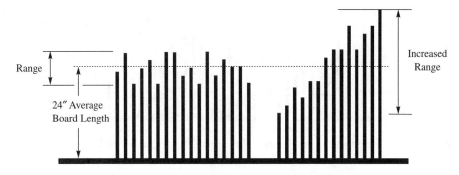

Figure 11.6 Stable Mean with Increased Range for Board Lengths

11.5   OUT-OF-CONTROL ACTIONS

Inevitably the question comes up, "Is the production line shut down if the control chart is out of control?" The answer to this question creates a dilemma for many firms. The simple answer is that *there must be corrective action to return the process to an in-control situation.* If there is no corrective action, then SPC adds no value and is therefore waste.

Acceptable corrective action is to stop production to correct an out-of-control process. On the surface, stopping production can be viewed as waste; however, defective product can not be built on a stopped line. With the proper team support, manufacturing resources quickly focus on problems, leading to line improvement. Over time, there is a reduction in production shutdowns.

Conversely, **line stops** with no corrective actions create continual production shutdowns, which lead to SPC disillusionment. People "look the other way" when the process is out of control to keep the line going. This includes blatant disregard for the SPC charts (which is common in many firms), or by clever tricks such as setting limits so wide that nearly all process conditions are in control. These practices are obviously wasteful.

The essential elements for an out-of-control SPC chart include appropriate corrective action to identify the special cause, correct the situation, and error-proof the problem so that it does not return. If the special cause is easily understood and resolved (e.g., resetting a valve or finding a fixture), then a minor line stoppage is done to correct the problem. Never just correct a simple problem—*make sure the problem does not happen again.* Otherwise, continually having minor line stops to correct the same trivial problem is waste.

An out-of-control situation may be more complicated to resolve. In a difficult case, the team may recommend running the out-of-control process while closely monitoring the output to ensure that product meets specification. This is done only until the special cause is identified and the process is brought back into control. This approach is acceptable, but it is easily abused and requires discipline to avoid letting

Equipment adjustments for out-of-control situations must be followed by error-proofing.
Photo courtesy of Ion Implant Services, Inc.

the temporary solution become permanent. Some firms practice this approach and may have a special name to alert operators that a process is out of control but still running within specification while corrective action is under way. The key is for the team to follow through and solve the special cause problem.

If process measurements are determined to be outside the specification limits (as seen in the subgroup data), then the situation is serious. *If parts are out of specification, then stop the line until the problem is corrected.* There is no alternative to this action. Short-term solutions can be implemented, such as an inspection to screen for defective parts until the final solution is implemented, but these actions are only short term. If a short-term solution turns into a long-term solution, then there is a fundamental problem that will only grow worse with time.

Stopping production at a workstation for any reason should not be taken lightly. Each team member must commit to doing everything possible to quickly resolve line shutdowns and return the line back to production status. Improvement will never come about by ignoring potential problems just to keep the line running. The key is to identify and correct the problems, small or large, and use error-proofing to keep the problem from returning. For a committed team, a line stop focuses resources on finding the root cause so that it does not return again.

Many firms accept minor line shutdowns because the production line is still capable of producing parts (due to WIP buildup at operations). If a process encounters frequent minor line stoppages, then this serves to justify more WIP buildup between operations. This approach is counterproductive, as the waste created by WIP is now being used to permit more waste (line stoppages), which means the problems are compounding. This is not competitive manufacturing.

We are now in what can be called "the gray area of manufacturing decision making." *There is no clear decision for solving the problem.* This happens because of the competing nature of the different manufacturing process technologies discussed in Chapter 5 (e.g., yield, cycle time, lot size, throughput). The optimum solution is a give and take between all these technologies, which requires technical analysis and judgment.

Manufacturing people previously looked outside of manufacturing to find solutions, which led to the growth and inefficiency of off-line manufacturing control. The most effective organization to solve manufacturing problems is a technically diverse, properly functioning team—a team capable of making "gray area" decisions to incrementally improve manufacturing.

✳ CASE STUDY ✳
Using SPC

A metal film layer is deposited on wafers in the metallization area of a wafer fab. This metal layer will be etched to form the circuit wiring between devices on the wafer. The technology used to apply the thin metal film is known as a PVD (physical vapor deposition), or sputtering. Sputtering takes place at a low pressure (in a vacuum), and is a process to remove metal from a source and uniformly apply it on a wafer.

This particular sputtering process occurs in a cluster tool, which has multiple process chambers (typically four) attached to the central tool body (the tool is shown in Figure 5.8). The wafer enters into a central transfer area of the cluster tool, and then is moved with a robot arm into process chambers depending on what process step is required.

For every lot of wafers (one cassette), there is at least one process monitor wafer included in the run. This process monitor wafer has been prepared to have the same wafer surface material as the product. When the run is complete, the operator removes the test monitor from the cassette and measures it for thickness using a measurement instrument.

After completing the test-monitor thickness measurements, the operator logs the lot into the floor control software system at the computer terminal. At that time, the system prompts for the test-monitor measurement values, and they are input into the software database.

For this software system, the SPC chart only comes up if there is an out-of-control situation. An operator can override the system and prompt the software to display the control chart. When the process is flagged for an out-of-control condition, the operator stops the process to take corrective action to understand why the process is out of control. The lot proceeds to the next process step.

Corrective action typically involves running another test monitor in the process chamber that created the out-of-control condition (a recent improvement permits the operator to track the test monitor by chamber). For the first rerun, the operator leaves all chamber conditions the same, because it is often the test monitor itself that creates the problem. It is hard to have good test monitors; although they have been cleaned for reuse, they often still have residues that affect the sputtering process.

If the test monitor fails again, then the operator makes additional runs while adjusting the process variable of time to compensate for the thickness. The operator is usually able to bring the tool back on line and sign it back into production. If the tool cannot be adjusted properly (which happens once every 6 months or so), then the operator will sign it down to maintenance.

Points to consider

1. Is SPC used properly to control this process?
2. What type of inspection is this? Is this inspection value or waste?
3. Is the test-monitor system acceptable or is it creating problems? What improvements could be made?
4. What are the implications of having the software display the chart only when it is out of control?

11.6 PROCESS IMPROVEMENT

Process improvement requires action by team members. *SPC charts only highlight a problem or potential problem.* SPC will never fix a problem—only people can do this. If a process is out of control and no person takes action to identify and correct the situation, then the out-of-control situation continues and the process degrades.

Unfortunately, some manufacturing firms choose to implement SPC in production, and then ignore the corrective action required to make improvements. It may appear bizarre that a firm would willingly neglect the potential to improve; however, *when a firm knowingly fails to improve, there are usually organizational or human interaction problems that inhibit technical improvements.* People are the reason for lack of improvement. You can now appreciate the importance of teams and their commitment to correcting problems to improve manufacturing.

A process improves if the team acts to identify, understand, and correct the special cause creating an out-of-control situation. The eight-step improvement plan is a framework for guiding the improvement activity.

In general, the nature of corrective action for improvement with SPC depends on whether the process has special cause or common cause variation. Let's look at how team corrective action differs for these two types of variation.

✷ CASE STUDY ✷
SPC as Waste

Note: This case study is not from a wafer-fab manufacturing line, but it is of-fered as a classic example of how SPC is used incorrectly. This situation is found in many firms.

The firm produces electronic control systems. The operators are informed that the company is implementing SPC. It will initially start at several opera-tions, and then expand plant wide. One of the initial operations is the fuse holder fabrication operation, which uses a high-speed punch press to make fuse holders.

Engineering and quality control (QC) are responsible for implementing SPC in production. The operators are given some information about what mea-surements to take and how to fill out the SPC charts. No training is given on interpreting the charts or on how to take corrective action to improve.

At the fuse holder operation, the throughput is 100 parts per minute. The operators are instructed by QC to check five parts every 15 minutes for SPC measurements without stopping the process. The parts require 1 minute to in-spect, and then 2 to 3 minutes of calculation and chart-plotting time. The QC engineer sets the upper and lower control limits at the same value as the spec-ification limits.

There are no changes made to the process at any time based on the SPC charts. The operators continue to make the same tool adjustments that were made prior to implementing SPC, which were based on observation and expe-rience. The only real difference is the additional time it takes to make the in-spections and fill out the SPC charts.

Periodically, a QC engineer comes and collects the charts. The operators are never informed about the charts or given any improvement information. This system continues for about a year, and then the plant is shut down and relo-cated to a foreign manufacturing site.

Points to consider

1. Why do you think SPC was started if there is no effort to improve the process?

2. If an operator thought that this SPC activity was not implemented properly, what options are available to make the SPC work correctly?

3. Would a team effort make a difference?

11.6.1 Special Cause Corrective Action

When a control chart is out of control, there is special cause variation. The objective is to *immediately* identify this source of variation and correct it so that the process returns to common cause variation. The process will then be in statistical control. This **special cause corrective action** corresponds to the cyclic effort of the eight-step improvement plan. There are various potential reasons for an out-of-control process, as shown in the section on control chart patterns. Some possible special causes include the following:

- Improper machine parameters, settings, or setup
- Improper measurement techniques
- Operator with insufficient training
- A change in the product material quality
- Tool wear or lacking adjustment

Special cause problems are often chronic loss, which are controlled through disciplined manufacturing and innovation, and are important because the first person to note an out-of-control situation is usually an operator. If the operator works within the team structure to potentially identify and correct the special cause problem without resorting to support from engineers and management, then this is efficient use of resources.

When an out-of-control situation occurs in production, the initial operator response should be to stop production and assess all available information to resolve the problem. The operator must take corrective action, which includes interpreting data (primarily SPC charts, test measurement results, source inspection data, and yield data) and evaluating the operation for any obvious cause (e.g., incorrect measurement, incorrect equipment settings, defective equipment, or incoming material that is out of specification). After solving the problem, verification tests are done to demonstrate that the special cause problem is corrected, followed by error-proofing. When the process is in control, the operator signs it back to production.

If the initial response uncovers a predictable chronic loss problem (such as tool misalignment that appears periodically), then it is corrected with minimal disruption to production. *Do not fall into the common trap of accepting chronic loss problems, because this leads to continual process disruption for the same problem.* Chronic loss problems must be investigated, corrected, and error-proofed—anything less is nonproductive work.

For some out-of-control situations, the special cause is not obvious. If a new problem is uncovered during the initial response, additional technical support may be required. The operator notifies a team member (e.g., the line technician, equipment technician) to analyze the problem and its potential impact to production. Until the root cause is identified and corrected, there should be no assumptions that the problem is minor and will go away by itself.

Always review individual data measurements on the SPC charts to determine if any data points are outside the specification limits. If so, then stop production to determine what action is necessary. Never produce defective parts.

11.6.2 Common Cause Corrective Action

A stable process operating with common cause variation is in statistical control and is a desirable situation. The goal is common cause variation because we can predict future process performance based on statistical probability. However, decision making is never so simple. The problem is that a stable process might not be optimal, which is contrary to the goals of competitive manufacturing.

To continually improve, the team must consciously *change a stable process into an unstable process to identify and correct new problems*. Taking **common cause corrective action** by recalculating and reducing the control limits will force improvement. Problems that were previously accepted as common cause variation now become special cause variation, requiring team corrective action.

To reduce control limits, recalculate the nominal target and upper and lower control limits with recent SPC data after the process has been stable for a period of time. If necessary, remove data points with special cause variation that have been corrected prior to recalculating the limits.

Reducing control limits is the team's primary tool for continual improvement, as shown in Figure 11.7. To interpret the chart, notice the process is stable with the initial control limits. Because the process is stable, the team recalculates the limits and reduces them to new levels. These are plotted on the chart as a new dashed line. Now

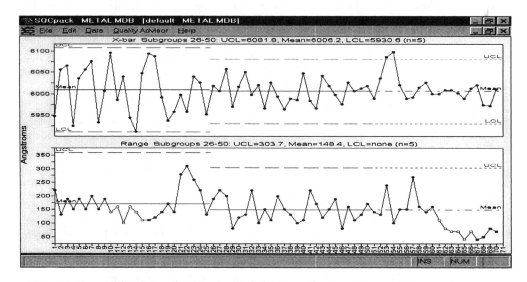

Figure 11.7 Reducing Control Limits with Common Cause Variation
SPC software courtesy of PQSystems, Inc. (www.pqsystems.com)

there are data points that are out of control with these new limits, which requires team corrective action. The chart would not have required action with the old limits.

After corrective action and improvement, the process restabilizes with the second set of limits. Once sufficient process stability is attained, the team again recalculates new limits, forcing them to act on problems that were previously accepted, and moves the process toward continual improvement. This movement corresponds to the linear effort necessary to move the process from platform to platform, *which is the way to guide the process toward improvement.*

 EXAMPLE 11.1 INTERPRETING A CONTROL CHART

Interpret the control chart in Figure 11.7. Explain the chart patterns, when the process was out of control, when it was stable, and why changes were made to the control chart limits.

Solution

The average chart is in control for the first twenty-five subgroups. The range chart is out of control due to a run of points below the nominal target. Even though this run of points on the range chart is goodness (more repeatability), it is important to identify the special cause to keep it from changing for the worse (which seems likely at data points 21 and 22). However, the team must have found and corrected the special cause for the range instability, because it returns to an in-control situation and remains that way. Because the special cause was identified, they decided to reduce the limits on the chart.

With the reduced limits, the process was stable for a relatively long period. There is an out-of-control situation for the average (exceeding the upper control limit at about subgroup data points 54 and 55). These data are out of control with the reduced control limits, yet probably would have been in control with the original control limits. The team takes corrective action and the average returns to an in-control situation.

It appears that the team has corrected a fundamental problem, because the average data become very repeatable (if this pattern continues, then the limits should be reduced again). The range is out of control for a run of points below the nominal target. There is a definite shift in the data, in the direction of goodness. The team must analyze the process to understand why this occurred so that they can maintain the process with this new level of variability. If this pattern continues, then the limits can be reduced again.

Each stable period on a control chart can be viewed as a process platform that requires cyclic team effort to iterate through the eight-step plan, identifying and correcting special causes. Control limits are fixed for this stable period, and do not

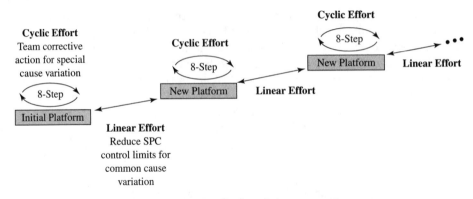

Figure 11.8 Process Platform Model with SPC for Continual Improvement

change until the process demonstrates stability with common cause variation. At this time, the process is ready for linear effort to force it into an unstable condition to uncover new problems.

You can now understand how SPC is directly linked to the eight-step improvement plan, shown in Figure 11.8. The process is stabilized by working within the iterative steps of the eight-step plan to eliminate special cause variation, which is cyclic effort. When stable, a process is moved to a new platform by reducing the control limits, which is linear effort. The team then continuously repeats the eight-step plan and reduces control limits, which is continual improvement.

11.6.3 Forcing a Process out of Control to Improve

An element of fear is associated with intentionally making a stable process go out of control. Why change a process when everything works right? Human nature is to accept an in-control process as good; however, a process cannot survive in a competitive market without improvement. Thus, we must manage personal fear in order to improve.

Change is not an isolated, abstract effort. Rather it is a logical extension of the present condition, as change implies a condition to change from. Fear of change is most easily managed by using the process platform model and SPC for improvement. A platform gives order to confusing information, providing a reference point for process knowledge. Confidence develops by acquiring knowledge about today's process, preparing the team to solve tomorrow's problems.

During continual improvement, a process alternates between in control and out of control, depending on where the process is on the process platform model. For this reason, *we cannot judge a process solely by whether it is in control. We can only judge a process based on the total team effort to improve.* If a team accepts defects and is not making the continual improvement effort, then an out-of-control situation is bad. However, if the out-of-control situation is part of the team's overall improvement activity within the process platform model, then it is a necessary part of the total effort to move a process toward optimum performance.

The difficulty of assessing whether a process should be in control or out of control is the reason why process control is the team's responsibility. Few people outside the process have sufficient technical knowledge to understand where a particular process is on the path to continual improvement.

Be aware that reducing the control limits of a stable process to achieve continual improvement is not practiced frequently in manufacturing firms. In fact, some firms do the opposite and open the control limits so that the process variable will never be out of control. Teams must consciously reject notions of goodness because a process is stable, and must educate management and other groups in the company about the benefits of intentionally forcing a stable process into an out-of-control situation, as this is the best way to be successful in a competitive market.

SUMMARY

SPC charts give feedback control for a manufacturing process. The three visual rules for assessing SPC are (1) the data plot within the control limits, (2) no runs on either side of the nominal target, and (3) no nonrandom patterns. Different control chart patterns are presented to assist in interpreting various patterns. Any chart that exhibits data that do not follow these three rules is deemed out of control, and corrective action is needed to identify and correct the special cause acting on the process. Two typical reasons for out-of-control processes are shifting mean or increased range, and analysis of these situations can help determine the source of the special cause variation.

The essential team response for an out-of-control situation is corrective action to bring the process back into control. This may involve a line stoppage to correct the special cause. Error-proofing is needed to avoid continual minor stoppages. Special cause corrective action responds to the ongoing problems in a process. Common cause corrective action forces a stable process out of control to achieve continual improvement.

IMPORTANT TERMS

Process control	Run of points
SPC feedback control	Nonrandom pattern
Control limits	Shifting mean
Nominal target	Increased range
Random pattern	Line stops
SPC visual rules	Special cause corrective action
Point outside of control limits	Common cause corrective action

REVIEW QUESTIONS

1. Describe the relevant questions about process control and capability addressed by SPC charts.
2. Draw a diagram and explain the concept of control for SPC.
3. What are the three basic indicators of control for an SPC chart?
4. State the three visual rules for assessing whether an SPC chart is in or out of control.
5. What is the key for success with the SPC rules?
6. Explain how a control chart could be out of control when the process is in control.
7. Explain how a control chart could have a run of points, but exhibit a positive trend.
8. If a control chart is out of control, is the manufacturing process shut down?
9. Explain the shifting average and the increasing range.
10. Discuss when SPC charts will correct a problem.
11. What action is done with special cause variation?
12. If product is found out of specification, what should be done?
13. What action is done with common cause variation?
14. How does an SPC chart lead a team toward continual improvement?
15. Why would a team force a stable process out of control?

EXERCISES

Interpreting Process Control Charts

1. Refer to the SPC chart for film thickness in Figure 10.8. Interpret this chart by answering the following questions:
   A. Is this chart in control for average and for range? Be specific.
   B. Is there any pattern that is occurring? If so, explain potential causes.
   C. The data are collected over 6 days. Break the data into 2-day intervals (8/1–8/2, 8/3–8/4, and 8/5–8/6). What can you say about the process performance for each of these three intervals? Explain in terms of both the average and the range.
2. Refer to the data and control chart in Exercise 3 from Chapter 10. Your team has continued collecting data for aluminum film thickness, which are shown on the following page. Plot these data on another control chart (label it Chart 2) using the same control limits and nominal targets from Exercise 3 in Chapter 10. Note: you may have to adjust the scale on the new chart.
   A. Is this process in control or out of control?
   B. Is it the average or range (or both) that is (are) changing?
   C. Why do you want the control chart limits to be fixed as you input these new data?

D. What should your team do now? Do you recommend stopping the process? (This is a critical product that must meet its ship date.)

Data for Wafer Aluminum Film Thickness, PVD Tool 1 (angstroms)

Date & Time	Test Site 1	Test Site 2	Test Site 3	Test Site 4	Test Site 5
3/16 8:30 A.M.	5,930	6,090	5,990	6,040	6,010
3/16 11:30 A.M.	5,940	6,000	5,890	5,950	5,990
3/16 4:30 P.M.	5,970	6,030	5,990	6,050	5,990
3/16 9:00 P.M.	5,830	5,900	5,880	5,860	5,840
3/17 2:30 A.M.	5,800	5,870	5,850	5,830	5,820
3/17 8:00 A.M.	5,990	5,930	6,050	5,980	6,060
3/17 10:30 A.M.	5,810	5,780	5,830	5,800	5,790
3/17 3:00 P.M.	5,870	5,800	5,910	5,920	5,930

3. Refer to the subgroup data presented in Exercise 4 of Chapter 10. Use this initial control chart and data in Chapter 10 as the control limits and nominal targets for the process. Now refer to the data below. Plot a new control chart with these data (label it Control Chart 2) using the original control limits and nominal targets. Note that the scale may change on this new chart.

A. Are the data in control or out of control? Be specific.
B. Explain how these data have changed relative to the initial set of data. Do you want to change the control limits when you input these new data?
C. Do these data meet specification (specification values are given in Chapter 10)?
D. Explain what team action should be taken for this process based on the control chart data.

Data for Sheet Resistance of Metal Film Layer, PVD Tool 5 (ohms per square)

Date & Time	Test Site 1	Test Site 2	Test Site 3	Test Site 4
8/13 8:30 A.M.	20.6	20.2	21.0	22.3
8/13 4:00 P.M.	22.0	21.2	21.3	22.4
8/14 6:30 A.M.	21.5	20.1	21.5	20.2
8/14 6:00 P.M.	18.7	19.6	21.0	20.1
8/15 5:00 A.M.	18.9	21.3	19.5	20.7
8/15 5:30 P.M.	18.6	19.8	20.9	19.4
8/16 6:30 A.M.	18.9	19.0	19.5	18.9
8/16 2:00 P.M.	18.5	19.8	20.4	19.0
8/17 5:30 A.M.	19.0	21.8	19.7	21.9
8/17 4:00 P.M.	18.2	19.4	19.8	20.0

4. A variables control chart is used to control a thin film metallization process. The variables datum is sheet resistance in ohms per square, which correlates directly to the thickness of the metal film (the thicker the film, then the lower the sheet resistance, and vice versa). The control limits and nominal target are indicated below, and are to be drawn on a variables SPC chart for $\bar{x}$ & R, using a suitable scale.

Sheet Resistance Control Limits Data (ohms per square)

UCL (average):	22.0 ohms/square
Nominal (average):	21.0 ohms/square
LCL (average):	20.0 ohms/square
UCL (range):	3.0 ohms/square
Nominal (range):	1.5 ohms/square
LCL (range):	0.0 ohms/square

Over the period of two shifts, you measure and record the following subgroup data on the SPC chart:

Resistance Measurement Subgroup Data (ohms per square)

1	2	3	4	5	6	7	8	9	10	11
21.1	19.4	21.2	21.6	21.1	19.8	21.9	22.7	22.8	22.8	20.0
21.7	20.9	21.5	21.3	20.4	21.2	22.7	22.1	22.9	21.3	22.9
20.4	21.7	20.5	21.1	20.7	21.9	21.9	21.9	22.8	22.5	21.3
20.8	19.1	20.9	21.8	21.3	22.1	22.5	22.6	21.9	21.4	20.6
21.0	20.7	21.0	21.0	20.9	20.0	22.8	21.5	22.3	20.9	21.8

A. Plot these subgroup data on the SPC chart with the control limits given above.

B. Identify which points are out of control by circling the points and explaining why.

5. Find a common everyday process, and describe it in terms of process control (examples are maintaining your weight, mowing the lawn, or studying for an exam). Identify how the following control concepts are used in the process:

A. Nominal target with acceptable randomness

B. Upper and lower control limits

C. How you know when control limits are exceeded

D. Potential special causes

6. The following control limits are in place for thickness measurements of an oxide film grown on a wafer. Put these limits on an SPC variables $\bar{x}$ & R control chart.

Wafer Film Thickness Control Limits Data (angstroms)

UCL (average):	1250 angstroms
Nominal (average):	1200 angstroms
LCL (average):	1150 angstroms
UCL (range):	100 angstroms
Nominal (range):	50 angstroms
LCL (range):	0 angstroms

Oxide thickness measurement data collected on wafers from different lots are:

Wafer Oxide Film Thickness Subgroup Data (angstroms)

1 (Subgroup):	1,200	1,190	1,210	1,200	1,170
2:	1,210	1,170	1,160	1,200	1,190
3:	1,180	1,210	1,170	1,190	1,160
4:	1,180	1,200	1,220	1,190	1,230
5:	1,250	1,220	1,190	1,190	1,180
6:	1,250	1,220	1,230	1,200	1,230
7:	1,190	1,170	1,200	1,200	1,180
8:	1,220	1,210	1,180	1,220	1,210
9:	1,220	1,170	1,200	1,210	1,190
10:	1,240	1,230	1,230	1,220	1,230
11:	1,220	1,230	1,210	1,220	1,200
12:	1,230	1,240	1,250	1,220	1,230
13:	1,260	1,210	1,240	1,250	1,250
14:	1,220	1,240	1,210	1,260	1,230
15:	1,250	1,260	1,250	1,260	1,230

A. Plot these thickness measurements on an SPC variables control chart.

B. Is there an out-of-control situation? If so, explain in what manner the chart is out of control.

7. Consider the following control chart trends. State the type of pattern and give at least three possible causes for the pattern.

Chart #1

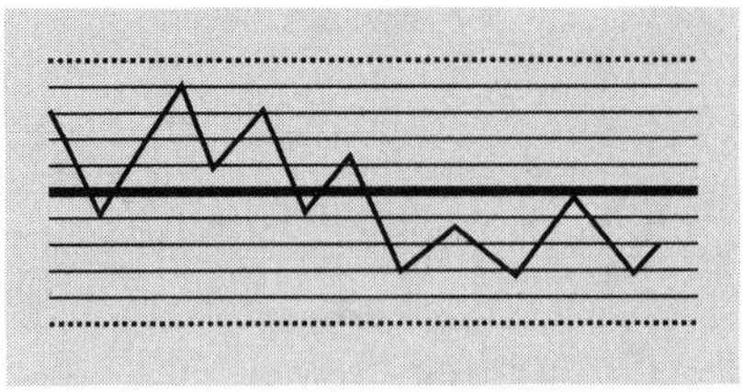

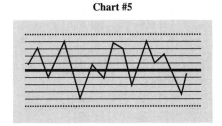

Chart #2

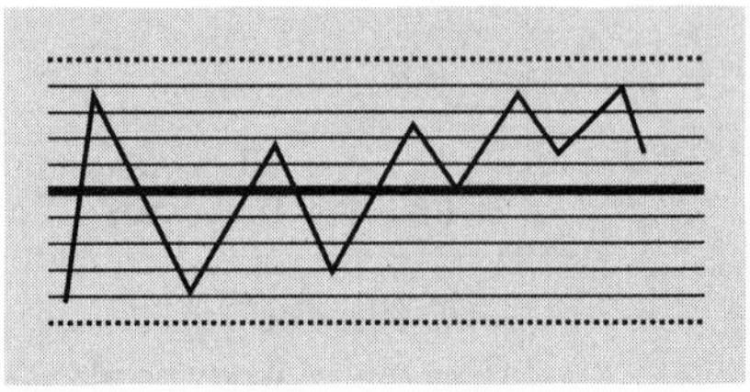

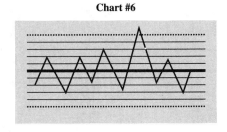

Chart #3

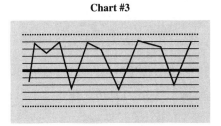

Chart #4

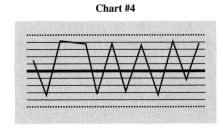

Chart #5

Chart #6

12

PROCESS CAPABILITY AND IMPROVEMENT

An in-control process is stable with predictable output. The risk is that the stable process is producing defective output, which is entirely possible when control of a process is demonstrated with data taken from the process itself. Through the use of process capability, operators can assess a process to verify that its controlled condition is acceptable for the specified requirements. This is a sanity check on the process and its performance.

After studying the material in this chapter, you should be able to:

1. Define *process capability*.
2. State the four conditions for process control and capability. Draw a normal curve of each situation, label the USL/LSL and UCL/LCL, and describe the appropriate team action for each condition.
3. When given the specification and process data, calculate and interpret C_p for a centered distribution. State when the process is incapable, capable, or robust.
4. When given the specification and process data, calculate C_{pk} for a noncentered distribution. State when the process is incapable, capable, or robust.

12.1 PROCESS CAPABILITY

Process capability estimates the ability of a process to produce parts that conform to the specification. Once a process is in control and incrementally improving, then capability demonstrates how good the process is against the ultimate requirement: the specification. If a process is capable, we have some measure of confidence that parts meet specification.

Process capability addresses the concern that the process could be statistically in control, but not capable of building parts to specification. How could this be? A process is in control when it produces measurements (the average and range) that statistically meet the requirements for process stability. However, recall that control limits are calculated from data that come from the process itself, *irrespective of the specification*.

If data from the process itself are the only criteria for being good, then we have a potential crisis, because the process could be producing bad parts yet be in statistical control. Capability is a means to "level-set" the process to ensure that parts meet the specification requirement. Different situations of process capability are shown in Figure 12.1, including how they would appear on a control chart. These four conditions for process capability are:

- Capable and in control
- Capable and out of control
- Incapable and in control
- Incapable and out of control

259

Case 1 Capable and In Control

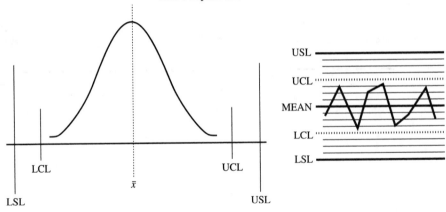

Case 2 Capable and Out of Control

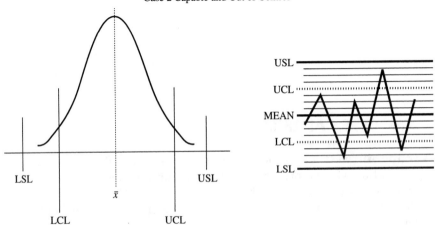

Case 3 Incapable and In Control

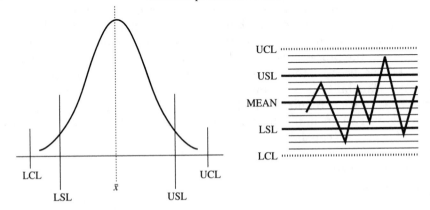

Figure 12.1 Different Process Capability Scenarios

260

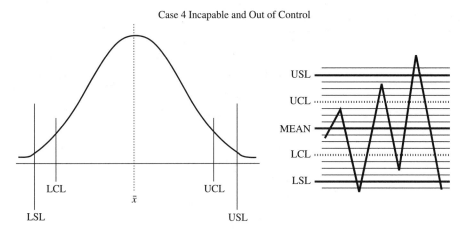

Figure 12.1 (continued)

You may suspect the optimum condition occurs if the process is **capable and in control.** This condition is desirable, but is not necessarily optimal. We cannot leave a process in this potentially suboptimum state. Based on what we have learned for achieving continual improvement, we know that even for an in-control process, it must be forced out of control to identify and correct new problems. The process is then **capable and out of control,** meaning it is unstable yet producing defect-free parts relative to the specification. Special cause problems are then identified, corrected, and error-proofed to bring the process back in control.

A dangerous situation is when a process is **incapable and in control.** In this case, the process is producing parts that are out of specification, even though the process is in control based on SPC data. Since SPC data are collected from the process irrespective of the specification limits, this situation can occur. Team action is required to identify and correct the root causes for the out-of-specification product while maintaining process stability. Production must be stopped until the team can ensure that all parts are within specification.

Another dangerous situation is a process that is **incapable and out of control.** This condition also requires that production be stopped. This type of situation might occur in the early stage of introducing a new product into manufacturing. The 8-step improvement plan should be used to introduce changes to the process. Note that neither of these incapable processes has a process window, and require inspection and corrective action until the process becomes capable and in control.

12.2 PROCESS CONTROL AND CAPABILITY

The four possible outcomes related to capability and control are listed in Table 12.1, including guidelines for team corrective action.

Table 12.1 Process Control and Capability

	Control	
Capability	Process is **in Control** (Statistically Stable)	Process is **out of Control** (Statistically Unstable)
Process is **capable** of meeting specification	• Healthy situation • Recalculate control limits • Make process unstable • Move to new process platform • 8-step plan to improve	• Requires action for special cause • Product is OK to ship • Implement corrective action • Restabilize process
Process is **incapable** of meeting specification	• Stop production • Potentially dangerous • Use source inspection to find defects • 8-step plan to improve • Monitor closely	• Stop production • Dangerous situation • Use team and 8-step plan • Major process changes required

The essential element for all four outcomes is the team's action. Without individuals working together as a team to correctly solve problems, the process degrades into an incapable and out-of-control process—complete disorder.

12.3   CALCULATING PROCESS CAPABILITY

Process capability is a useful concept that can be numerically calculated as a **capability index.** This capability index has become widely used in industry as an indicator of process performance against the specification tolerance. As with all endeavors, a fair assessment of process capability depends on properly collected data that represent the process conditions.

To calculate the index, we will numerically compare the specification tolerance to the process tolerance. We will not use control limits to calculate capability, but instead use the calculated sigma to represent the process tolerance. For our calculations, we will use a $\pm 3\sigma$ variation to represent the process, which can be changed to higher sigma levels for a more severe test of capability.

12.3.1 General Process Capability

The general formula for the process capability index is:

$$\text{Capability Index } (C_p) = \frac{\text{Specification Tolerance}}{\text{Process Tolerance } (\pm 3\sigma)}$$

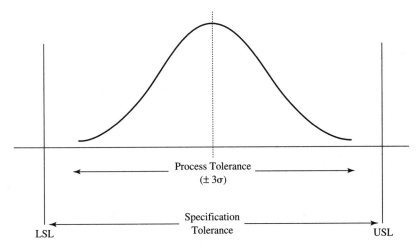

Figure 12.2 Capable Process with Process Window

where

Specification tolerance = Upper Spec Limit – Lower Spec Limit

Process tolerance ($\pm 3\sigma$) = 2 × (3σ) (Note: this represents both tails
of the process.)

This formula is simply a ratio of the specification tolerance over the process tolerance. The higher this number when calculated, the more capable the process. Give some thought about this ratio. If the specification limits are large relative to the process limits, then the capability index is large. This situation is good because it now has a wide process window to produce parts without going out of specification.

The goal in manufacturing is to have as wide a process window as possible to build parts, thus reducing scrap and rejects (which is the opposite of product development's goal, which is to have as narrow a process window as possible to protect the product's performance). Figure 12.2 shows a capable process, with a wide process window, whereas Figure 12.3 shows an incapable process with no process window.

In years past, during the capacity production era, manufacturing firms had little control over their process. *Because manufacturing could neither control nor reduce the process tolerance, the only choice for increasing process capability was to coerce product development in to make the specification tolerance wider.* This was known as "opening up the spec." This approach is a poor solution to improvement, because it creates cross-functional fighting, leads to politicized equipment, and will eventually reduce product performance (as tolerances are wider).

With team action and continual improvement activities, process capability is increased by reducing the process tolerance (improving $\pm 3\sigma$ to larger sigma values, such as $\pm 4\sigma$ and beyond). This method puts manufacturing teams in charge of their process, and contributes to the firm's competitive market position by permitting products to be designed with tighter specification tolerances.

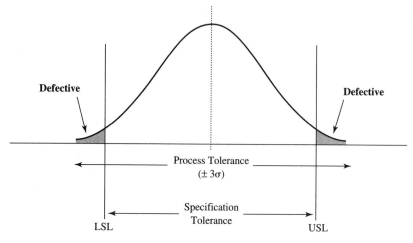

Figure 12.3 Incapable Process with No Process Window

12.3.2 Calculating Process Capability—Centered Distribution

Let's consider the situation where the process is centered within the specification, as shown in Figures 12.2 and 12.3. In practice, a process is evaluated for centrality by comparing the mean of the sample data with the nominal specification value. If the two are equal or nearly equal, then the process is centered in the specification.

To calculate the process capability for a **centered process** distribution, C_p, use the following formula:

$$C_p = \frac{USL - LSL}{\pm 3\sigma}$$

where

$$C_p = \text{process capability for a process distribution centered in the specification tolerance (the sample average is close to the nominal dimension).}$$

$USL - LSL$ = specification tolerance

$\pm 3\sigma = 2\,(3\sigma)$ = process tolerance

Process capability index has no units, since it is a ratio. Process capability improves as C_p increases. Table 12.2 shows the interpretation for different C_p values.

If C_p is less than 1, as shown in Figure 12.4, then we have an **incapable process** and will produce defective parts. A **capable process** always has a C_p value greater than 1, as in Figure 12.5. A **barely capable** process has a C_p of 1. The general industry guideline for C_p values is to achieve a $C_p > 2.0$. When the C_p is greater than 2.0, the process is termed **robust,** since the specification tolerance is twice the process tolerance.

Table 12.2 Interpretation of C_p Values

C_p Value	Capability	Reason
$C_p < 1$:	Incapable Process	Specification Tolerance < Process Tolerance
$C_p \approx 1$:	Barely Capable	Specification Tolerance $\approx$ Process Tolerance
$C_p > 1$:	Capable Process	Specification Tolerance > Process Tolerance
Industry process capability goal: $C_p > 2$		

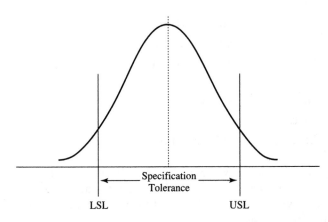

Figure 12.4 $C_p < 1$ (C_p is approximately 0.75)

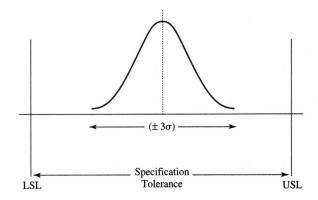

Figure 12.5 $C_p > 1$ (C_p is approximately 2)

---

* **EXAMPLE 12.1 PROCESS CAPABILITY—CENTERED DISTRIBUTION**

Calculate C_p using the following data collected for wafer film thickness:

Average ($\bar{x}$): 396 angstroms
Sigma (σ): 5 angstroms

To illustrate C_p, use two different specification limits:

Case I:	Nominal:	400 angstroms
	USL:	420 angstroms
	LSL:	380 angstroms
Case II:	Nominal:	400 angstroms
	USL:	410 angstroms
	LSL:	390 angstroms

Calculate the process capability for each case. (Note: in manufacturing, there are never two specifications for one process. View Case I and Case II as two separate processes.) Explain why the capability is different for Case I and Case II. Is the process robust in either case?

Solution

For Case I:

$$C_p = \frac{USL - LSL}{2 \times 3(\sigma)} = \frac{420 - 380}{2 \times 3(5)} = \frac{40}{30} = 1.33 \; (capable)$$

This process is capable, and has a process window; however, it is not robust (C_p needs to be greater than 2 to be robust). Process improvement should continue.

For Case II:

$$C_p = \frac{USL - LSL}{2 \times 3(\sigma)} = \frac{410 - 390}{2 \times 3(5)} = \frac{20}{30} = 0.66 \; (incapable)$$

This process is not capable because C_p is less than 1 (it has no process window), and is obviously not robust. The process must undergo basic improvement, and team action is required to control the quality of the parts until the process becomes capable. Capability is lower for Case II because of the reduced specification tolerance.

You might notice that the C_p ratio does not take into account where the mean of the process is located relative to the nominal specification dimension. It is insensitive to whether the process distribution is closer to the USL or LSL, which is dangerous, as parts are rejected if they are beyond the specification limits in either direction. Therefore, we will modify the capability index formula to take into account whether one side of the process is ready to go out of specification. We will term this new capability index the C_{pk} for a noncentered distribution.

12.3.3 Calculating Process Capability—Noncentered Distribution

In many process conditions, the process average is not centered directly on the nominal dimension of the specification, which reflects that the mean of the sample rarely equals the nominal dimension of the specification (although it can occur for an improved process). The capability index for a **noncentered process** is termed C_{pk}. For C_{pk}, we are interested only in the distribution tail on the side where the process average ($\bar{x}$) is nearest the specification limit. This is the particular specification limit where the part has the highest probability of going beyond the limit. In the case of Figure 12.6, this would be the tail closest to the lower specification limit (LSL).

An analogy for C_{pk} is driving a car on a freeway with construction barriers on each side of the road, as shown in Figure 12.7. If the car starts veering to the right side of the road, our concern is hitting the right side of the car. This is where we focus our attention to avoid hitting the concrete barriers. It would be illogical to also worry about the concrete barriers on the left side, if the car is ready to hit the barriers on the right side.

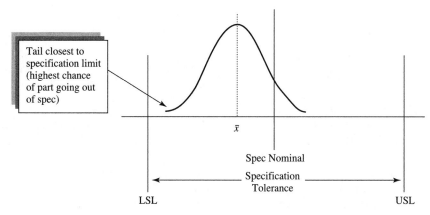

Figure 12.6 Noncentered Distribution

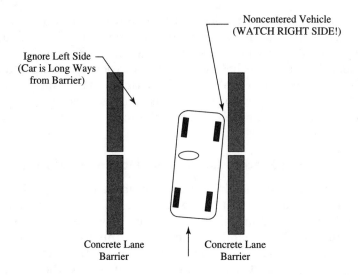

Figure 12.7 Noncentered Vehicle with Risk of Hitting One Side

To calculate the process capability for a noncentered process, C_{pk}, the formula is:

$$C_{pk} = \frac{\text{Minimum of}\,(\,|\bar{x} - USL|\ \text{or}\ |\bar{x} - LSL|\,)}{3\sigma}$$

where

$\quad\quad\quad\quad C_{pk}$ = process capability for noncentered process

$\quad\quad\quad\quad \bar{x}$ = process mean (obtain from a sample or the nominal target of the control chart)

$(\,|\bar{x} - USL|\ \text{or}\ |\bar{x} - LSL|\,)$ = choose LSL or USL to obtain the minimum value from $\bar{x}$.

$\quad\quad\quad\quad 3\sigma$ = we are only considering one side of the distribution; therefore use 3σ, which is the variability in one tail.

The C_{pk} process capability index is sensitive to processes with shifted means, and reflects that the product is more likely to be out of specification at the process tail closest to the specification limit. Figure 12.6 highlights this situation. Table 12.3 describes how to interpret different values of C_{pk} for process capability.

C_{pk} is similar to C_p in that a capable process has a $C_{pk} > 1$, as illustrated in Figure 12.8. C_{pk} for an incapable process is shown in Figure 12.9. Because it is a conservative estimate of process capability, it is more difficult to attain a C_{pk} index as large as C_p. The industry goal for a robust process is $C_{pk} > 1.5$.

C_{pk} is the best estimate of the process capability. It is conservative because it only considers the distribution tail that is closest to the specification limit. If the

Table 12.3 Interpretation of C_{pk} Values

C_{pk} *Value*	*Capability*	*Reason*
$C_{pk} < 1$:	Incapable Process	Specification Limit < Worst Case Process Tolerance
$C_{pk} \approx 1$:	Barely Capable	Specification Limit $\approx$ Worst Case Process Tolerance
$C_{pk} > 1$:	Capable Process	Specification Limit > Worst Case Process Tolerance
Industry process capability goal: $C_{pk} > 1.5$		

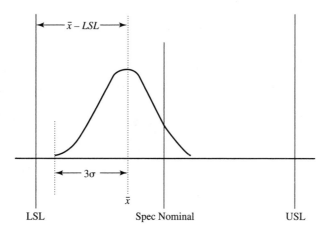

Figure 12.8 $C_{pk} > 1$ (C_{pk} is approximately 1.25)

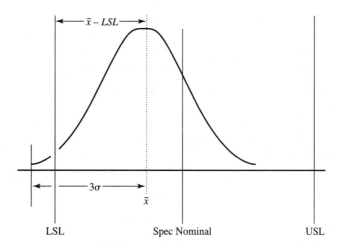

Figure 12.9 $C_{pk} < 1$ (C_{pk} is approximately 0.9)

process distribution is not centered on the nominal specification dimension, use C_{pk}. If the process distribution happens to be centered, then $C_{pk} = C_p$. In any other situation, C_{pk} will be less than C_p.

 EXAMPLE 12.2 PROCESS CAPABILITY—NONCENTERED DISTRIBUTION

Calculate C_{pk} using the following data for wafer film thickness:

Average ($\bar{x}$): 398 angstroms

Sigma (σ): 8 angstroms

To illustrate how C_{pk} can change, we will use two different specification limits. (Note: in manufacturing, there is only one specification at an operation; we are considering two as an example. View these two cases as two separate processes.)

Case I: Nominal: 400 angstroms   Case II: Nominal: 385 angstroms
         USL: 430 angstroms   USL: 415 angstroms
         LSL: 370 angstroms   LSL: 355 angstroms

Solution

Case I:

$$C_{pk} = \frac{|\bar{x} - LSL|}{3\sigma} = \frac{|398 - 370|}{3(8)} = \frac{28}{24} = 1.17 \; (capable)$$

Case II:

$$C_{pk} = \frac{|\bar{x} - USL|}{3\sigma} = \frac{|398 - 415|}{3(8)} = \frac{17}{24} = 0.58 \; (incapable)$$

Case I is capable of producing parts to specification, but improvement is important because the process is not robust. The process for Case II is incapable. It requires fundamental improvement (use the eight-step plan), and also requires special team effort to ensure no defective parts escape the process. This is done through source inspection and rework until the process improves and is capable.

12.3.4 Process Capability and Improvement

For a new, unstable process, we expect the $C_{pk} < 1$. It is the nature of a new process to be unstable. Through use of the eight-step improvement plan, we improve

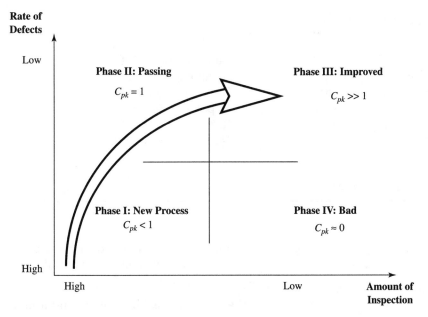

Figure 12.10 Process Improvement and C_{pk}

the process and move to $C_{pk} > 1$. Let's reconsider the process inspection efficiency model using C_{pk}, as shown in Figure 12.10.

The benefit of process capability is its ability to measure manufacturing process performance relative to the specification. As design ground rules tighten for products in competitive markets, capability analysis provides confidence that teams can meet reduced specification criteria in manufacturing. This confidence permits the product design group to specify tighter product specifications with the knowledge that manufacturing is capable. In this manner, both manufacturing process capability and improvement position a company to provide the most advanced product in the marketplace while meeting its competitive goals of lowest cost, highest quality, and shortest delivery.

SUMMARY

Process capability measures how a process performs relative to the specification. There are different team responses depending on whether a process is in control or out of control, and capable or incapable. A process capability index is calculated through C_p for a centered process and C_{pk} for a noncentered process, and is used to assess capability relative to the specification.

IMPORTANT TERMS

Process capability	Centered process C_p
Capable and in control	Incapable process
Capable and out of control	Barely capable
Incapable and in control	Capable process
Incapable and out of control	Robust
Capability index	Noncentered process C_{pk}

REVIEW QUESTIONS

1. What is process capability?
2. Explain the four following process conditions: capable and in control, capable and out of control, incapable and in control, and incapable and out of control.
3. What is the general formula for the process capability index?
4. What is the relationship between the process capability index and a manufacturing process window?
5. State and explain the formula for calculating process capability with a centered distribution.
6. For what C_p values is a process incapable, barely capable, capable, and robust?
7. Draw a normal curve with specification limits for an incapable, barely capable, and capable process.
8. State and explain the formula for calculating process capability with a noncentered distribution.
9. For what C_{pk} values is a process incapable, barely capable, capable, and robust?
10. Draw a sketch showing how C_{pk} changes with respect to defects and inspection as a process improves.

EXERCISES

Process Capability

1. The thickness specification for an aluminum metal layer on a wafer is:

    USL: 6200 angstroms
    Nominal: 6000 angstroms
    LSL: 5800 angstroms

An operator performs 200 thickness measurements using a thickness-measuring device on a sample of four test wafers from a run, and finds the following sample data:

Mean ($\bar{x}$) = 5940 angstroms
Sigma (σ) = 50 angstroms

A. Calculate the C_p and C_{pk} for this process based on the above specification and measurement data.
B. Is the process incapable, capable, or robust?

2. The specification limits for resist thickness after spin-coating are:

USL: 2.8 microns
Nominal: 2.0 microns
LSL: 1.2 microns

You take five measurements on three wafers and obtain the following data:

Data for Resist Thickness (microns)

Wafer 1:	2.2	1.8	2.4	2.3	2.2
Wafer 2:	2.2	1.7	1.9	1.9	1.9
Wafer 3:	2.3	2.4	2.1	2.1	2.2

A. What is the C_p for these data? (Hint: first calculate the mean and sigma for the data.)
B. Is the process incapable, capable, or robust?

3. For the same resist thickness data in Exercise 2, an improved resist is implemented in production to attain smaller critical dimensions. The new resist thickness specification is:

USL: 2.4 microns
Nominal: 2.0 microns
LSL: 1.6 microns

A. With this new specification, what is the C_p given the existing resist thickness data that were measured in Exercise 2?
B. Is the process incapable, capable, or robust?
C. What would be a reasonable goal for improvement activity?

4. To analyze the capability of a new setup procedure to control the pressure of the contact pins for electrical test, the process engineer and the lead technician carefully set up the equipment and process parameters by developing a carefully

documented procedure. They then run twenty parts to measure the pressure, and obtain the following:

**Pin Contact Pressure
Data (grams force)**

n = 20 measurements	
15	15
16	15
15	14
14	15
15	16
16	15
15	15
15	14
16	15
15	15

The specification for the contact pressure is:

USL: 20 grams force
Nominal: 15 grams force
LSL: 10 grams force

A. Calculate C_p and C_{pk}. Is this process incapable, capable, or robust?
B. Do you recommend implementing this new setup procedure for controlling pin contact pressure?

5. The new equipment setup procedure for controlling pin contact pressure in Exercise 4 has been on the floor for 6 months, and used on all shifts. A random measurement of testers and their contact pin pressure results in the following data:

**Pin Contact Pressure
Data (grams force)**

n = 10 measurements	
19	16
6	21
18	16
10	19
15	8

    A. Based on these data, is this process incapable, capable, or robust?

    B. If there is a problem with capability, what would you assess as the source of this problem (based on the information given in this exercise and the previous exercise)?

6. You are a field representative for an equipment supplier. While on a service call, a customer claims that your tool keeps producing bad parts. You suspect the tool is acceptable and it may be a problem with incoming parts from another operation. Your customer agrees to let you make some sheet resistance measurements on incoming parts. You sample parts and find a mean of 30 ohms and a sigma of 1.5 ohms.

    For your company's tool to work right, you know that the incoming parts must be within a specification of 30 +/–3 ohms. You have to call your supervisor in 20 minutes and explain the problem.

    A. What is the C_p for the incoming parts?

    B. Are the incoming parts from a capable process?

    C. Could there be out-of-specification parts?

✳ APPENDICES

Appendix 1   Summary of Different SPC Charts
Appendix 2   Explanation of SPC Chart Patterns and Possible Causes
Appendix 3   Standard Normal Probability Table
Appendix 4   Variables Statistical Process Control Chart
Appendix 5   Wafer Fab Terminology
Appendix 6   Flowchart of Team Actions for Low OEE
Appendix 7   Glossary

APPENDIX 1

✳ SUMMARY OF DIFFERENT SPC CHARTS

VARIABLES DATA

Variables data charts create a picture of a process variable over time, allowing us to separate the sources of variation into common and special cause for effective control actions.

The types of variables data control charts summarized are:

- $\bar{x}$ & S
- *Individuals and Moving Range*
- *Moving Average and Moving Range*
- *Run Chart*
- *Median*

$\bar{x}$ & S

The $\bar{x}$ & S (standard deviation) is a variation of the $\bar{x}$ & R chart. The $\bar{x}$ chart (top chart in Figure A1.1) shows the average of each subgroup data. Instead of range, the lower chart in Figure A1.1 shows the standard deviation (shown as S, σ_{n-1}, or σ' on some calculators) of each subgroup. This control chart has become common in manufacturing because SPC software or hand calculators can calculate the standard deviation with ease. It should always be used when the subgroup size has ten or more data points.

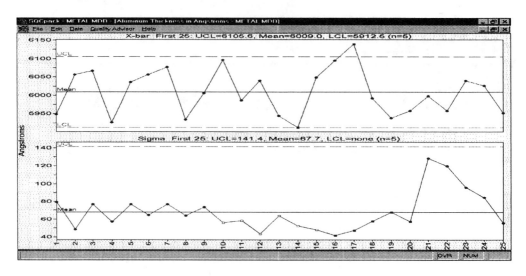

Figure A1.1 *x-bar & S* Plot

SPC software courtesy of PQSystems, Inc. (www.pqsystems.com)

Individuals and Moving Range

The individuals (X) and moving range (MR) chart (Figure A1.2) is used with sub-groups that only contain one reading. This occurs sometimes in manufacturing due to a limitation on the amount of data that can be in a subgroup. The X chart, on top, shows each individual reading. The MR chart, on the bottom, creates ranges by finding the difference between consecutive subgroups. It uses the absolute value to avoid negative moving range values.

Moving Average and Moving Range

Moving averages work much the same as moving ranges work on the individuals and moving range chart. It is used with subgroups containing one reading, but instead of plotting the individual values, plot an average of two or more readings (as defined for the specific chart), shown in Figure A1.3. This will smooth out minor variations between individual readings and will show how the system is running over time.

Run Chart

A run chart (Figure A1.4) is a line graph of individual data points plotted over time. It is used as a preliminary analysis of data to look for patterns or trends due to special cause variation. The run chart is sometimes useful in manufacturing because it is simple and can quickly show how a variable is changing over time. An example would be monitoring the particle count in a tool over time to control cleanliness during a run.

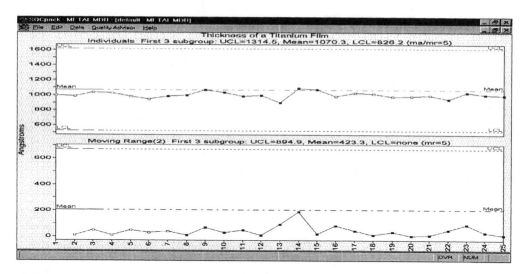

Figure A1.2 Individuals and Moving Range Plot
SPC software courtesy of PQSystems, Inc. (www.pqsystems.com)

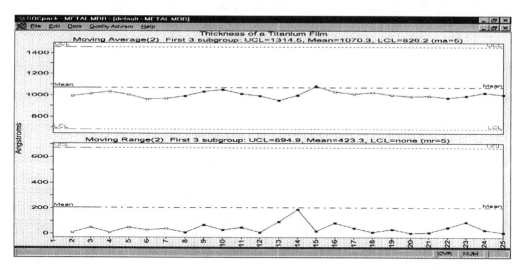

Figure A1.3 Moving Average and Moving Range Plot
SPC software courtesy of PQSystems, Inc. (www.pqsystems.com)

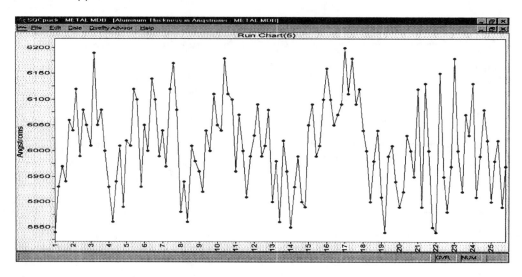

Figure A1.4 Run Chart
SPC software courtesy of PQSystems, Inc. (www.pqsystems.com)

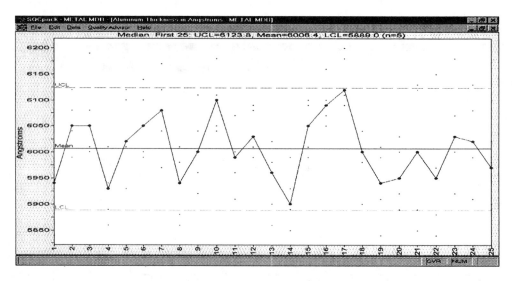

Figure A1.5 Median Plot
SPC software courtesy of PQSystems, Inc. (www.pqsystems.com)

Median

Median charts use the median (or middle) value of a subgroup to monitor process stability, as shown in Figure A1.5. Since sigma is often used to analyze data in manufacturing, this plot is not too common. It does permit the user to show all the individual data points on the chart.

ATTRIBUTE DATA

Attribute data can be classified and counted (e.g., number of good and bad parts). There are two different types of attribute data:

Nonconforming data: A count of defective parts (e.g., good/bad, tall/short, or pass/fail)

Nonconformities data: The number of defective occurrences that should not be present but are (e.g., dents, scratches, bubbles, and missing buttons). In this case, the product does not necessarily fail, but it has defects associated with it.

There are some areas of manufacturing, such as test operations, that collect and analyze extensive attribute data. In this case, attribute charts could be useful. An example is the test area in wafer fabs known as wafer sort, which tests individual die on a wafer for electrical performance.

Use attribute control charts carefully in manufacturing to ensure that defective parts are not accepted in the process (avoid thinking, "I made defects, but since I did not exceed the control limits, I'm ok"). It is imperative that the team establishes a control limit reduction plan to reduce the defects to zero.

The four basic types of attribute charts used in manufacturing are:

- *np-charts*
- *p-charts*
- *c-charts*
- *u-charts*

Attribute data charts are categorized based on whether the sample is fixed or varied, and whether the data are nonconforming or nonconformities, as shown in Table A1.1.

np-charts

The np-chart is used to monitor the number of nonconforming units (defective units) within a subgroup that is always the same size. The np-chart shows the actual

Table A1.1 Attribute Control Chart Categories

	Defective Units (Nonconforming)	A Count of Defects (Nonconformities)
Fixed Sample Size	np	c
Variable Sample Size	p	u

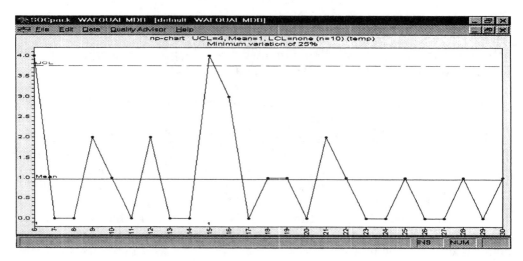

Figure A1.6 np Attribute Chart
SPC software courtesy of PQSystems, Inc. (www.pqsystems.com)

number of defective units in each subgroup, since the subgroup size is constant. An np control chart is shown in Figure A1.6.

p-charts

The p-chart uses a proportion of the nonconforming (or defective) items because the subgroup size may vary (since sample size can affect the probability of finding a defect). The size of the subgroup is typically large to help find the nonconforming parts. Generally, the subgroup size should not vary by more than +/– 25 percent. If this occurs, then the control limits are adjusted to compensate. A p attribute chart is shown in Figure A1.7.

The data needed to calculate the proportion nonconforming are (1) the total number of parts inspected in a subgroup, and (2) the number of nonconforming parts found.

c-charts

The c-chart (Figure A1.8) measures the number of nonconformities (discrepancies, or count of defects) in an inspection lot (as opposed to the number of units found nonconforming, as plotted on a p-chart). Examples of nonconformities are scratches, stains, or bubbles. For c-charts, the subgroup size must be constant. This chart shows the actual number of nonconformities per subgroup.

u-charts

The u-chart measures the number of nonconformities (discrepancies) per inspection reporting unit in subgroups, which can have varying sample sizes (or amounts of material inspected). A u attribute chart is shown in Figure A1.9.

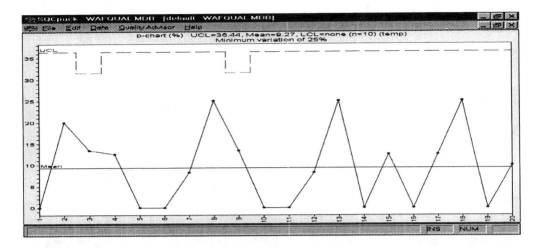

Figure A1.7 p Attribute Chart
SPC software courtesy of PQSystems, Inc. (www.pqsystems.com)

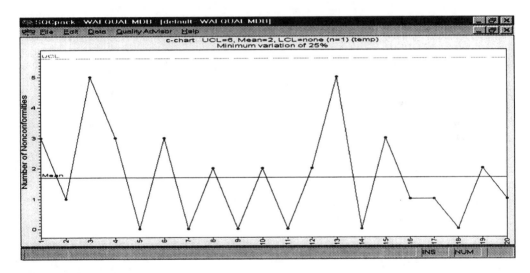

Figure A1.8 c Attribute Chart
SPC software courtesy of PQSystems, Inc. (www.pqsystems.com)

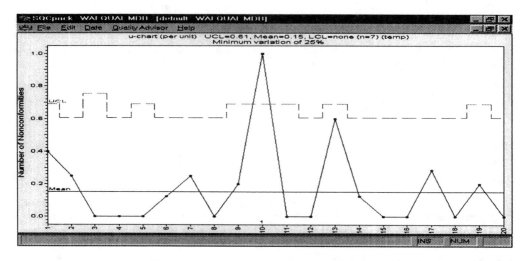

Figure A1.9 u Attribute Chart
SPC software courtesy of PQSystems, Inc. (www.pqsystems.com)

APPENDIX 2

✳ EXPLANATION OF SPC CHART PATTERNS AND POSSIBLE CAUSES[1]

Cycles Short trends in the data that occur in repeated patterns.

Average   Seasonal effects such as temperature and humidity.
Worn positions or threads.
Operator fatigue.
Rotation of people on the job.
Difference in measurement tools.
Voltage fluctuations.
Regular difference between shifts.

Range   Maintenance schedules.
Operator fatigue.
Rotation of fixtures or gauges.
Wear of tool or die (causing excessive play).

Freaks Presence of a single data point that differs greatly from the others for no explainable cause.

---

[1]See Reference 3 in the bibliography.

Average   Freaks do not ordinarily show up on an average plot without a cor-
responding indication on the range plot.
Wrong setting, corrected immediately.
Error in measurement or plotting.
Incomplete or omitted operation.

Range   Accidental damage in handling.
Error in measurement or plotting.
Incomplete or omitted operation.
Setup parts.
Some obvious physical abnormality which can be detected by exam-
ining the units in the sample that produced the freak point.

Gradual Change in Level   Some element in the process that affects a few units
at first and then more and more as time passes.

Average   Gradual introduction of new material.
Improved skills of operator.
Change in maintenance program.
Introduction of process controls in this or other areas.

Range   Change in fixtures.
Change in methods.
Improved skills of operator.

Grouping or Bunching   Measurements cluster together in a nonrandom fashion.

Average   Measurement difficulties.
Change in the calibration of a test set or measuring instrument.
Different person making the measurements.

Range   Freaks in the data.
Mixture of distributions.

Instability   Pattern exhibits unnaturally large fluctuations.

Average   Overadjustment of a machine.
Fixtures or holders not holding the work in position.
Carelessness during setup.
Mix of different lots of material.
Differences in test sets or gauges.
Erratic equipment behavior due to software or hardware.
Effect of screening and sorting operations at various steps in the
process.

Range   Untrained operator.
Excessive play in equipment or fixture.

Machine in need of repair.
Mixture of material.
Unstable testing equipment.

Interaction   Tendency of one variable to alter the behavior of another variable.

Average   Relationship between operator skill, training, tool performance, design of tool, and batches of material.

Range   Relationship between same variables.

Sudden Shift in Level   Abrupt change in the data points in one direction.

Average   Change to a new kind of material or lot of material.
New operator or inspector.
New test part or monitor.
New machine, machine setting, setup, or method.

Range   Change in motivation of operators.
New operators.
New equipment.
Change to a different material or different supplier of material.

Systematic   Pattern becomes predictable (e.g., a high point always follows low point).

Average   Difference between shifts.
Difference between test sets or product monitors.
Difference between assembly lines where product is sampled in rotation.
Systematic manner of dividing the data.

Range   Effect is usually due to a systematic manner of dividing the data.
Less frequently there may be a large difference in spread between different conveyors, shifts, sources of material, and so forth, being sampled in rotation.

Trends   A continuous movement either up or down, or a long series of points without a change in direction.

Average   Tool wear.
Wear of threads, holding devices, or gauges.
Deterioration of chemical solution.
Inadequate maintenance on test set or monitor.
Seasonal effects, including temperature and humidity.
Human variables.
Operator fatigue (for manual operation).

Increases or decreases in production schedules.
Gradual change in standards.
Poor maintenance or housekeeping procedures of equipment (e.g., accumulation of dirt and clogging of filters).
Pumps becoming dirty.

Range   Something loosening or wearing gradually.
Wearing of a tool.
Various types of mixtures.
Effect of changed maintenance program.

APPENDIX 3

STANDARD NORMAL PROBABILITY TABLE

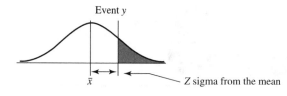

Event y

Z sigma from the mean

$$Z = \frac{|y - \bar{x}|}{\sigma}$$

Z	x.x0	x.x1	x.x2	x.x3	x.x4
4.0	0.00003				
3.9	0.00005	0.00005	0.0004	0.0004	0.0004
3.8	0.00007	0.00007	0.00007	0.00006	0.00006
3.7	0.00011	0.00010	0.00010	0.00010	0.00009
3.6	0.00016	0.00015	0.00015	0.00014	0.0001
3.5	0.00023	0.00022	0.00022	0.00021	0.0002
3.4	0.00034	0.00032	0.00031	0.0003	0.00029
3.3	0.00048	0.00047	0.00045	0.00043	0.00042
3.2	0.00069	0.00066	0.00064	0.00062	0.00060
3.1	0.00097	0.00094	0.00090	0.00087	0.00084
3.0	0.00135	0.00131	0.00126	0.00122	0.00118
2.9	0.0019	0.0018	0.0018	0.0017	0.0016
2.8	0.0026	0.0025	0.0024	0.0023	0.0023
2.7	0.0035	0.0034	0.0033	0.0032	0.0031
2.6	0.0047	0.0045	0.0044	0.0043	0.0041
2.5	0.0062	0.0060	0.0059	0.0057	0.0055
2.4	0.0082	0.0080	0.0078	0.0075	0.0073
2.3	0.0107	0.0104	0.0102	0.0099	0.0096
2.2	0.0139	0.0136	0.0132	0.0129	0.0125
2.1	0.0179	0.0174	0.0170	0.0166	0.0162
2.0	0.0228	0.0222	0.0217	0.0212	0.0207
1.9	0.0287	0.0281	0.0274	0.0268	0.0262
1.8	0.0359	0.0351	0.0344	0.0336	0.0329
1.7	0.0446	0.0436	0.0427	0.0418	0.0409
1.6	0.0548	0.0537	0.0526	0.0516	0.0505
1.5	0.0668	0.0655	0.0643	0.0630	0.0618
1.4	0.0808	0.0793	0.0778	0.0764	0.0749
1.3	0.0968	0.0951	0.0934	0.0918	0.0901
1.2	0.1151	0.1131	0.1112	0.1093	0.1075
1.1	0.1357	0.1335	0.1314	0.1292	0.1271
1.0	0.1587	0.1562	0.1539	0.1545	0.1492
0.9	0.1841	0.1814	0.1788	0.1762	0.1736
0.8	0.2119	0.2090	0.2061	0.2033	0.2005
0.7	0.2420	0.2389	0.2358	0.2327	0.2297
0.6	0.2743	0.2709	0.2676	0.2643	0.2611
0.5	0.3085	0.3050	0.3015	0.2981	0.2946
0.4	0.3446	0.3409	0.3372	0.3336	0.3300
0.3	0.3821	0.3783	0.3745	0.3707	0.3669
0.2	0.4207	0.4168	0.4129	0.4090	0.4052
0.1	0.4602	0.4562	0.4522	0.4483	0.4443
0.0	0.5000	0.4960	0.4920	0.4880	0.4840

Note: This is for a single-tail distribution. For a double-tail (both sides of the normal curve), look up each Z value independently, and then add them for the total probability.

292

z	x.x5	x.x6	x.x7	x.x8	x.x9
4.0					
3.9	0.00004	0.00004	0.00004	0.00003	0.00003
3.8	0.00006	0.00006	0.00005	0.00005	0.00005
3.7	0.00009	0.00008	0.00008	0.00008	0.00008
3.6	0.00013	0.00013	0.00012	0.00012	0.00011
3.5	0.00019	0.00019	0.00018	0.00017	0.00017
3.4	0.00028	0.00027	0.00026	0.00025	0.00024
3.3	0.00040	0.00039	0.00038	0.00036	0.00035
3.2	0.00058	0.00056	0.00054	0.00052	0.00050
3.1	0.00082	0.00079	0.00076	0.00074	0.00071
3.0	0.00114	0.00111	0.00107	0.00104	0.00100
2.9	0.0016	0.0015	0.0015	0.0014	0.0014
2.8	0.0022	0.0021	0.0021	0.0020	0.0019
2.7	0.0030	0.0029	0.0028	0.0027	0.0026
2.6	0.0040	0.0039	0.0038	0.0037	0.0036
2.5	0.0054	0.0052	0.0051	0.0049	0.0048
2.4	0.0071	0.0069	0.0068	0.0066	0.0064
2.3	0.0094	0.0091	0.0089	0.0087	0.0084
2.2	0.0122	0.0119	0.0116	0.0113	0.0110
2.1	0.0158	0.0154	0.0150	0.0146	0.0143
2.0	0.0202	0.0197	0.0192	0.0188	0.0183
1.9	0.0256	0.0250	0.0244	0.0239	0.0233
1.8	0.0322	0.0314	0.0307	0.0301	0.0294
1.7	0.0401	0.0392	0.0384	0.0375	0.0367
1.6	0.0495	0.0485	0.0475	0.0465	0.0455
1.5	0.0606	0.0594	0.0582	0.0571	0.0559
1.4	0.0735	0.0721	0.0708	0.0694	0.0681
1.3	0.0885	0.0869	0.0853	0.0838	0.0823
1.2	0.1056	0.1038	0.1020	0.1003	0.0985
1.1	0.1251	0.1230	0.1210	0.1190	0.1170
1.0	0.1469	0.1446	0.1423	0.1401	0.1379
0.9	0.1711	0.1685	0.1660	0.1635	0.1611
0.8	0.1977	0.1949	0.1922	0.1894	0.1867
0.7	0.2266	0.2236	0.2206	0.2177	0.2148
0.6	0.2578	0.2546	0.2514	0.2483	0.2451
0.5	0.2912	0.2877	0.2843	0.2810	0.2776
0.4	0.3264	0.3228	0.3192	0.3156	0.3121
0.3	0.3632	0.3594	0.3557	0.3520	0.3483
0.2	0.4013	0.3974	0.3936	0.3897	0.3859
0.1	0.4404	0.4364	0.4325	0.4286	0.4247
0.0	0.4801	0.4761	0.4721	0.4681	0.4641

APPENDIX 4

VARIABLES STATISTICAL PROCESS CONTROL CHART

Variables Control Chart : $\bar{X}$ & R (Average & Range)

Part Number						Operation										Specification					Chart Number			
Parameter						Machine										Subgroup Size					Unit of Measure			
Operator																								
Date																								
Time																								
Subgroup Data 1																								
2																								
3																								
4																								
5																								
Average $\bar{X}$																								
Range R																								
	1	2	3	4	5	6	7	8	9	10	11	12	13	14	15	16	17	18	19	20	21	22	23	24	25

Average ($\bar{X}$)

Range (R)

295

APPENDIX 5

✳ WAFER FAB TERMINOLOGY

A general process flow for a complementary metal oxide semiconductor (CMOS) in a wafer fab is shown in Figure A5.1 and discussed in this appendix.

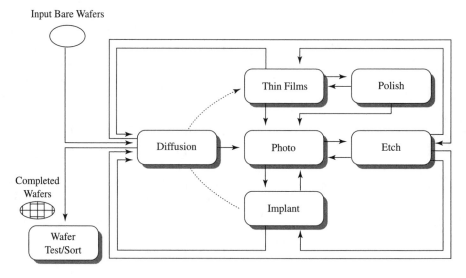

Figure A5.1 General CMOS Process Flow[2]

---

[2]See Reference 13 in the bibliography.

WAFER AND PROCESS TECHNOLOGY

Cassette: The standardized carrier for wafers in a wafer fab. Cassettes hold twenty-five wafers and have a standard dimension for interfacing with different tools.

Chip: One of the individual integrated circuit (IC) devices on a wafer. Also referred to as *die*.

Clean Room: An area in which semiconductor fabrication takes place. The cleanliness of the room is highly controlled to limit the number of particles to which the semiconductor is exposed.

Cluster Tool: An integrated, computer-controlled workstation that has several process chambers clustered around a vacuum transfer chamber and input-output section.

CMOS: Complementary metal oxide semiconductor. A particular type of chip technology that has n-channel and p-channel MOS transistors on the same chip.

Critical Dimension: A reference to the smallest dimension on a wafer, which is down to 0.18 to 0.25 micron width of lines in a modern wafer fab. Wafer fabs control their process to ensure that critical dimensions are acceptable.

Die: Another term for the individual chip on a wafer.

Layers: Refers to the process of adding or growing thin layers of material to the wafer surface. The different materials are selectively removed through an etch process.

Metallization: Application of a thin metal layer to the wafer surface, to be used as circuit wiring between the devices. Metallization is typically followed by photo and etch to make the circuit patterns.

Planarization: A flat-surfaced (planar) structure fabricated by adding or growing layers.

Resistivity: A measure of the resistance to current flow in a material.

Silicon: The most common material for fabricating semiconductors. It is used because of its special atomic structure that makes it a semiconductor material.

Submicron: Refers to wafer fab technology with a critical dimension of less than 1 micron.

Thin Films: A term describing the growing or depositing of thin layers of material on a surface. This could also be termed *metallization* if it is a metal layer.

Wafer: A thin, round slice of typically silicon semiconductor material from which chips are made.

PHOTOLITHOGRAPHY

Align and Expose: Tool to align the photomask to the correct location on a resist-coated wafer. After alignment, light-sensitive photoresist on the wafer is exposed by light passing through nonopaque areas of the photomask.

Develop: Chemical process that sprays liquid developer on the patterned resist on the wafer surface and dissolves the exposed resist. The result is a visible pattern of circuitry on the wafer surface.

Exposure: Method of defining circuit patterns by the interaction of light or other forms of energy with a photoresist material that is sensitive to the energy source.

Hot Plate: Method to heat a single wafer to improve adhesion, curing, etc.

Patterning: Another term for photolithography.

Photolithography: The process to transfer an image from a photomask to the light-sensitive resist material that is on the wafer surface, using light to effect the transfer.

Photomask: Quartz or glass plate that has the fine pattern of circuits and devices (the master pattern) used in photolithography. There is a photomask for each layer on the wafer. Also termed *reticle* or *mask*.

Photoresist: Light-sensitive liquid chemical (also referred to as *resist*) that is coated on the wafer to be used as the photographic medium for the transfer of a circuit pattern from the photomask to the wafer.

Resist Coat: The operation of coating a thin liquid resist on the wafer during a spin process.

Reticle: see Photomask.

Surface Preparation: Cleaning and drying the wafer surface to prepare for a process step.

ETCH

Etch: The process step in chip manufacturing that uses chemicals or a dry process to selectively remove layers of a material from the surface of the wafer. This forms the circuitry and devices on the different layers.

Etch Rate: The rate (speed) that etching occurs in an etch tool.

Plasma Etch: A dry etch process using reactive gases energized by a plasma field.

THIN FILMS

Metallization: Deposits a thin metal layer on the wafer surface.

Thin Film: The term for a thin layer on the wafer surface. This layer could be metal or nonmetal.

ION IMPLANT

Concentration: The amount of dopants in the silicon. It is typically measured as resistance, since dopants are ions (have a positive or negative charge) and will change the resistance of silicon depending on their concentration.

Dopant: An element that alters the conductivity of a semiconductor by contributing either a hole or electron to the conduction process. For silicon, the most common dopants are boron (p-type or hole as majority carrier) and phosphorus (n-type or electron as majority carrier).

Ion: An atom that has either gained or lost electrons, making it a charged particle (either negative or positive).

Ion Implantation: A process that injects (implants) an ionized atom such as boron into a semiconductor wafer.

DIFFUSION

Anneal: A high-temperature processing step designed to reduce stresses in the crystal structure of the wafer.

Diffusion: A process in semiconductor manufacturing of introducing small amounts of impurities (dopants) into a substrate material, such as silicon. The furnace process is often referred to as *diffusion*.

Furnace: A piece of equipment used to uniformly heat up the wafer in a controlled environment.

WAFER SORT

Alignment Target: A geometric shape on the wafer used for precisely positioning the wafer during processing.

Probe Card: Has the fine diameter pins that will contact the pads on each individual die for testing.

Probe Pins: The fine diameter pins that contact the die for testing.

Tester: A tool that performs electrical test on wafers to verify that each individual die is acceptable.

Wafer Sort: The test operation where each individual die on a wafer is electrically tested for function and performance. Bad die have an ink drop placed on them or are stored in a computer database for rejection once the die are sliced from the wafer.

MEASUREMENT TECHNOLOGY

Angstroms: Unit that represents 10^{-10} meters. There are 10,000 angstroms in a micron.

Four-Point Probe: Precise resistance measurement procedure using two points to apply a current and two points to measure a voltage. It is used to determine the sheet resistance of a thin film.

Micron: Metric unit that represents 10^{-6} meters, or one-millionth of a meter.

Ohms per square: Unit of measurement for sheet resistance that correlates the thickness of a thin film to the resistance reading.

Sheet Resistance: A resistance measurement with units of ohms per square centimeter that shows the number of p-type and n-type donor atoms in a semiconductor or determines the thickness of a thin film.

APPENDIX **6**

※ FLOWCHART OF TEAM ACTIONS FOR LOW OEE[3]

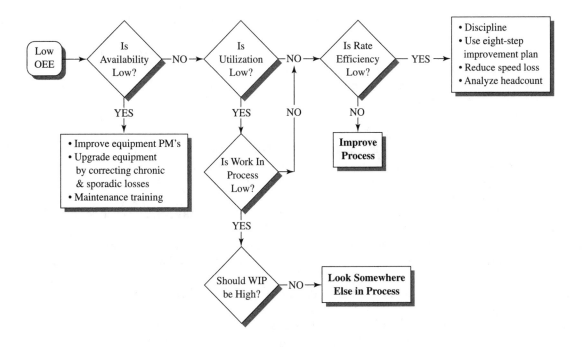

[3]See Reference 14 in the bibliography.

APPENDIX 7

✳ GLOSSARY

5S: A quality improvement program that originated in Japan and establishes discipline in manufacturing.

6 Sigma: A quality improvement program that uses the statistical variation of a population to represent quality. It assumes all process means shift 1.5 sigma, which leads to overly optimistic estimates for the number of defects associated with each sigma level.

Acceptance Quality Level (AQL): A statistical technique of assessing the number of defects in a sample to determine if the sample has fewer defects than a predetermined level, thus permitting defects to pass into or out of a process.

Accuracy: The deviation of a measured value from the true value.

Assignable Cause: Special cause variation in a process with an assignable reason for its occurrence.

Attribute Data: Qualitative data assigned to a part that can be counted for recording and analysis.

Automated Equipment: Manufacturing equipment that requires minimal human intervention to process parts by using integrated software control, sensors, chemical delivery, and robotic handling.

Availability: Amount of time manufacturing equipment is available for production.

Average: The sum of all measured values divided by the number of measurements (i.e., sample size). Designated by the symbol $\bar{x}$, the average is the unique number that best represents the central tendency of the measured data.

Balanced Line: Production that is evened out with respect to cycle time, which serves to reduce process WIP.

Batch Processing: The processing of parts in large groups.

Bathtub Curve: A model of equipment reliability performance with an initial phase of early failures, followed by a stable phase of low failures, and eventually a wearout period of increased failures.

Bell-Shaped Curve: A common description for the normal probability distribution.

Biased Data: The collection of data that misrepresents the intended population.

Bimodal Process: Two distinct normal population distributions that often exhibit two separate peaks.

Black Box Production: A strategy of minimizing a firm's effort to manufacture product. Raw materials are input into the process and finished goods are output.

Bottleneck: The slowest operation in a manufacturing process that defines the throughput of the process.

Brainstorming: An idea-generation technique based on an open exchange of team ideas to seek the optimum solution to a problem.

C_p: An index that represents how capable a process is of producing in-specification product if the process is centered in the specification. It is defined as the specification tolerance divided by the process tolerance.

C_{pk}: An index that represents how capable a process is of producing product in specification when the process is not centered within the specification. It addresses the distribution tail on the side where the process average is nearest the specification limit.

Capability Index: The specification tolerance divided by the $+/- 3$-sigma process tolerance.

Capable Process: A process that can repeatedly produce product that meets specification.

Capacity Production: A manufacturing strategy of building the maximum number of product possible, often to the detriment of efficiency and quality.

Cause-and-Effect Diagram: A problem-solving technique that uses a graphic description to find the root cause of a problem.

Central Tendency: The centralizing of normal data around a certain value with increasing sample size.

Chronic Equipment Losses: Small, frequent equipment failures that occur regularly in production, to the point that we accept them as normal.

Common Cause Corrective Action: The action necessary to force a stable process into instability by reducing the control limits, permitting the team to identify and correct new special cause variation.

Common Cause Variation: Describes the natural behavior of any process. The variation of the process occurs randomly in a predictable manner.

Competitive Manufacturing: The ability to manufacture products for lowest cost, highest quality, and shortest delivery.

Complementary Technical Skills: A team with all necessary technical skills represented to assist in problem solving.

Constraint: A location in a process flow that restricts product throughput. Also referred to as a *bottleneck*.

Continual Improvement: The philosophy that manufacturing must increase value and continually improve by reducing waste.

Control Chart: A graphical representation of a process variable that statistically analyzes common and special cause variation to make decisions about process performance.

Control Limit Reduction: The practice of reducing control limits on a stable control chart to highlight new special cause problems for continual improvement.

Control Limits: Lines on a control chart referred to as the upper and lower control limits that are used to judge the significance of the variation from subgroup to subgroup.

Corrective Action: A controlled change to a process to correct a problem.

Corrective Maintenance: Effort to support the existing equipment conditions in a process with no improvement relative to the current process methods.

Cycle Time: The time allotted to make one piece or unit in production. It can be defined for a specific process step or multiple processes.

Cyclic Effort: The effort expended to correct day-to-day problems to achieve a stable process.

Data: Numerical information collected from a small number of parts using a sampling plan and often used to make predictions.

Discipline: The need for people working in manufacturing to adhere to work procedures and practices to ensure repeatability in the process.

Distribution: A method to describe the measured data from a population by showing how the individual data points are related. The normal distribution is represented by the bell-shaped curve and defined by the average and standard deviation.

Downstream Operation: Any operation in a process flow that follows a preceding operation.

Economic Order Quantity: A manufacturing production control model that emphasizes large lot sizes and determines how to most efficiently hold inventory.

Economies of Scale: A production model that attempts to maximize the quantity of parts moving through a process.

Eight-Step Improvement Plan: A structured continual improvement plan that iterates through eight steps to address improvement.

Elements of Manufacturing: The infrastructure of resources required to produce a product: people, methods, machines, and materials.

Equipment Capacity: The amount of product processed per unit time through equipment at a workstation.

Error-Proofing: A philosophy of ensuring that a corrected problem does not return.

Fishbone Diagram: A visual tool that identifies all potential causes for a problem in a picture format. Also known as cause-and-effect diagram.

Floor Control System: The method for controlling the flow of different products through a manufacturing process.

Floor Layout: The manner that equipment is grouped on a manufacturing floor that defines how product will flow through the process.

Flowcharting: A visual description of a process from start to finish by defining the flow from step to step.

Formal Team: A team created by management that addresses a specific temporary task or exists as an ongoing team.

Frequency of Occurrence: The number of times a particular data point occurs in a sample population. This is often near the average of the population.

Good Data: Collected data that correctly represent the total population.

High-Volume Manufacturing: Manufacturing associated with continuous product flow, standardized work procedures, part interchangeability, and repeatability of the process.

Histogram: A visual plot of data that is categorized into intervals to show how individual data points are related.

Holistic Process View: A comprehensive view of all the interdependencies in a manufacturing process.

Idle Time: A waste in a manufacturing process that occurs when a resource is not being used to add value to the product.

Incapable Process: A manufacturing line that is not able to build product that repeatedly meets the specification.

In-Control Process: A process that has only common cause or natural variation.

Increased Range: The range of a measured variable increases while the mean remains stable.

Incremental Improvement: Process improvement that makes small process changes based on current knowledge.

Informal Team: A group of employees working together to address a production need or solve a problem.

Inspection: The effort to verify that the product conforms to the specification requirements.

ISO 9000: A quality certification procedure that primarily addresses documentation control, calibration, and general quality principles through self-assessments and audits.

Just-In-Time (JIT) Manufacturing: A method to optimize product flow in a high-volume production line by delivering the necessary parts at an operation only when they are needed.

Kanbans: A flow communication system in high-volume manufacturing that indicates and controls the number of parts between operations.

Large Sample: A sampling technique to obtain sufficient data to properly represent the total population. It is often limited by economic considerations.

Linear Effort: The long-term work necessary in a process to uncover previously accepted problems. It moves a process linearly toward continual improvement.

Line Stops: The stopping of a manufacturing line due to unanticipated production problems.

Lot Size: The grouping of parts into a particular quantity, or batch, for processing. This may hide process problems by permitting work at operations even when there is a problem.

Low-Volume Manufacturing: Manufacturing based on discontinuous product flow and associated with prototype work or early product development. It is also referred to as *job shop production.*

Lower Control Limit (LCL): The minimum value for common cause variation on a control chart. Points that plot outside of this limit exhibit special cause variation.

Lower Specification Limit (LSL): The minimum permissible specification value for a product variable. Product built below this limit must be scrapped or reworked.

Malcolm Baldrige National Quality Award: A U.S. quality award that is given yearly to companies that demonstrate quality practices. The award is based on self-assessment, with an on-site inspection for finalists in defined categories.

Manufacturing: The process of adding value to a material to build a product, typically involving a repetitive sequence of process steps.

Manufacturing Data: Measurements collected to provide numerical data about process variables.

Manufacturing Efficiency: An indication of value added to a product during manufacturing, with efficient manufacturing only expending the necessary effort to produce the product.

Manufacturing Improvements: Effort expended in manufacturing to increase the value added during production by reducing waste.

Manufacturing Process: A sequence of operations used to build a product.

Manufacturing Productivity: The effort to manufacture products within the shortest delivery time defined by the market needs.

Manufacturing Quality: The manufactured product conforms to all specification requirements.

Market: Products compete in a market. The market conditions can determine whether a manufacturer is competitive or not.

Material Requirements Planning (MRP): An inventory control method that estimates the amount of material needed to build product to the published production schedule.

Mean: The sum of all individual values divided by the total number of values. It is the unique number that best represents a population. See *average*.

Mean Time between Failure (MTBF): The average time a tool runs between breakdowns.

Mean Time to Repair (MTTR): The average time it takes to repair a tool.

Measurement Error: Data that do not represent the population due to reasons such as incorrectly calibrated measurement tools.

Median: The middle value in a group of measurements, when arranged from the lowest to the highest.

Nominal Dimension: The ideal value for each specified parameter.

Nominal Target: The location for central tendency on a variables control chart, which is x-double bar for the average plot and R-bar for the range plot.

Nonproductive Work: Unnecessary effort expended in a manufacturing process.

Nonrandom Pattern: An obvious pattern that does not exhibit randomness on a control chart, indicating special cause variation.

Normal Curve: A continuous, symmetrical, bell-shaped frequency distribution for variables data. It is described by the mean and standard deviation, and is the basis for statistical process control charts.

Normal Probability Distribution: An equivalent term for the normal curve.

Off-Line Manufacturing Control: A form of manufacturing management where decisions are made outside of manufacturing away from the process.

On/Off Test: A method of determining the most likely cause of a problem by assessing whether it can make the problem appear and disappear (turn on and off).

Operator: The person who has the equipment operation and process knowledge to produce product at a workstation.

Operator Cross-Training: A manufacturing program of training operators to work at multiple workstations for flexibility.

Outcome: Each possible result of an event.

Out-of-Control Actions: The team actions necessary to respond to an out-of-control manufacturing situation.

Out-of-Control Process: A process that has special cause (unnatural) variation present. This process is unstable and requires corrective action.

Overall Equipment Effectiveness (OEE): A diagnostic tool for analyzing equipment and process performance for improvement.

Overproduction: Manufacturing waste caused by the production of product not required by customer demand.

Parallel Work: An effort to perform multiple tasks in parallel to reduce the cycle time at an operation.

Pareto Chart: A method of analyzing various problems by ranking them in order of importance to prioritize corrective action.

Part Interchangeability: The ability to mix the same part in a high-volume process because it is manufactured to nominal dimensions with tolerances.

Population: The collective group to be measured, and exists as a total or sample population.

PPM (parts per million): Describes the number of defective parts per million produced.

Precision: Indicates a measurement's repeatability with respect to a mean value of a sample.

Preventive Maintenance: Equipment support activity to correct problems.

Probability: The chance that a certain outcome will occur, given all possible outcomes.

Process Capability: An estimate of how capable the process is to produce parts that repeatedly meet the specification.

Process Control: The manufacturing effort to control process variables to produce a predictable product output, usually with statistical process control (SPC).

Process Flow: The unique set of manufacturing steps that define how a product is produced.

Process-Oriented Layout: A manufacturing floor layout that groups similar tool technologies together.

Process Platform: A structured improvement method that resolves current problems in a process prior to introducing change.

Process Stability: A process that functions under only common cause variation.

Process Window: The relationship between the specification tolerance and the process tolerance. Ideally, the specification tolerance is larger than the process tolerance.

Product: The outcome of a manufacturing process. Products compete in a marketplace for customers.

Product Defect: A part that does not conform to specification and can be scrapped or reworked.

Product-Oriented Layout: A manufacturing floor layout that groups together the tools necessary to build a product.

Pull Process: A manufacturing process flow method that "pulls" parts through a process based on customer demand.

Push Process: A manufacturing process flow method that "pushes" parts through a process irrespective of the demand at each operation.

Quality Circles: A team-based improvement activity that involves all people associated with manufacturing in the improvement process.

Quantum Leap Changes: A risky form of improvement that involves revolutionary change with little relationship to the existing process knowledge.

Random Pattern: The desirable pattern for how points plot about the nominal target on a control chart.

Random Sample: Every member of the population has an equal chance of being selected for measurement in the sample group.

Range: The difference between the highest and lowest values in a sample, indicating the amount of spread in the measured data.

Research and Development (R&D): Scientific investigation at the limits of technology to find new methods and applications for knowledge.

Reverse Engineering: The practice of analyzing a product to understand how it works and then determining a better way to design and build it.

Rework: The manufacturing effort to correct defective parts.

Root Cause: The true cause of a problem.

Run of Points: A group of data points (usually seven or more) on an SPC chart that plots on one side of the nominal target.

Sampling Plan: A method of obtaining statistical data by defining the procedure for how the data are collected.

Scrap Parts: Parts manufactured out of specification. They are usually thrown away.

Series Work: The act of doing work activities one step after another (serially). This typically requires the longest time to perform a task.

Setup Time: The time it takes to change over equipment to run different products in manufacturing.

Shifting Mean: The sudden change in the mean of a measured variable while the variability remains stable.

Sigma: A numerical description for data dispersion about the mean for the total population. In practice, sigma is often used to describe data variability for both the sample and total population.

Sigma Level: The sigma level of a process defines the number of acceptable and defective parts produced.

Single-Piece Processing: A manufacturing process that moves product through an operation a single piece at a time, which often reduces the operation cycle time.

Single-Variable Change: A method of introducing change in manufacturing by varying only one process variable at a time.

Source Inspection: The philosophy of controlling quality at the location in the process where the defect can occur.

Special Cause Corrective Action: The action necessary to identify and correct special cause variation in a process.

Special Cause Variation: The unnatural behavior of a process. It is nonrandom and has an assignable cause.

Specifications: The documented engineering requirements that define acceptability of product and process parameters.

Sporadic Losses: Major equipment failures that are corrected with a maintenance team that has special equipment training.

Stable Process: A process that is in statistical control (only common cause variation).

Standard Deviation: A numerical description for how data are dispersed about the mean in a sample population.

Standardization: The replication of work procedures at the same operation from different equipment and people.

Station Yield: The number of good product exiting a workstation divided by the total number of product that entered the workstation.

Statistical Process Control (SPC): The use of statistical techniques for analyzing a process to predict future performance based on common and special cause variation.

Statistics: The collection, analysis, and interpretation of numerical data that represent a population.

Subgroups: The most fundamental group of data that is collected, measured, and analyzed on control charts.

Takt Time: The rhythm of a process represented by the time interval between each product produced. A balanced line will have a regular takt time.

Team: A group of people working together with a common mission to meet routine production goals, complete projects, or solve problems.

Team Attributes: Characteristics associated with teams that describe how team members work together.

Team Members: Individuals in a manufacturing team that represent the technical skills needed in production.

Technician: The person responsible for troubleshooting and maintaining the equipment and process in manufacturing.

Throughput: The actual quantity of parts processed per unit time in a manufacturing process.

Tolerance: The acceptable range of dimensions for producing a part in manufacturing. It defines the upper specification limit (USL) and lower specification limit (LSL).

Total Customer Satisfaction (TCS): A philosophy that the customer must be satisfied in order to successfully complete a task.

Total Preventive Maintenance (TPM): An equipment maintenance system developed in Japan that strives for optimum equipment performance in manufacturing.

Total Process Yield: A multiplication of the yield of each workstation in a total process to represent the true yield of a process.

Total Productive Management (TPM): A manufacturing improvement program that assembles many different improvement activities into one comprehensive program.

Total Quality Control: A term first used in the 1950s in the United States and expanded by Japanese manufacturers to address quality practices at all levels of the manufacturing process.

Total Quality Management (TQM): A management tool for implementing quality improvement practices throughout an organization.

Traditional Organization: The most common form of manufacturing management with a department of operators reporting to a supervisor and a hierarchy of managers above the supervisor.

Training: A method to transfer knowledge from the subject matter experts to people working in manufacturing.

Unstable Process: A process with special cause (unnatural) variation present.

Upper Control Limit (UCL): The upper limit on a control chart that defines common cause variation. Points that plot outside the control limit exhibit special cause variation.

Upper Specification Limit (USL): The maximum permissible value for specified product criteria. Product manufactured with any criteria outside of the upper specification limit is unacceptable.

Upstream Operation: The operation that precedes a following operation in a process flow.

Value: The effort added in manufacturing to transform a material into some functional form (or product) that is useful to humans.

Variability: A measurement of how data are dispersed about a mean.

Variable: A process parameter that can be varied between different settings based on the needs of the process to produce a part.

Variables Data: Quantitative data from variables where measurements are used for analysis.

Variation: The inevitable differences among individual measurements of a population. The sources of variation can be grouped into two major categories: common cause and special cause.

Visual Flow Test: Assessing a process for logical product flow to assess efficiency.

Waste: Effort expended in manufacturing that is not necessary to build the product.

Work in Process (WIP): Represents the product stored at the various steps in the manufacturing line, with excessive WIP indicating process inefficiency.

Workstation: The location on the manufacturing floor where work occurs in a process.

$\bar{x}$ and R Chart: A variables process control chart for the average and range.

$\bar{x}$ and S Chart: A variables process control chart for the average and standard deviation.

Yield: The ability of a manufacturing process to produce product with acceptable quality. Production yield is the number of good product divided by the total number of product started.

Z Statistic: A method of calculating the probability of an event based on knowing how many sigma the event is from the mean.

Zero Defects: A manufacturing philosophy that accepts no defects in manufacturing.

✳ BIBLIOGRAPHY

1. Ames, V.A., and Powell, David. 1996. *TPM Workbook*. Austin, TX: Sematech.
2. Amrine, Harold T., Ritchey, John A., and Hulley, Oliver S. 1982. *Manufacturing Organization and Management*, 4th Edition. New Jersey: Prentice-Hall.
3. AT&T. 1956. *Statistical Quality Control Handbook*. Charlotte, NC: Delmar Printing Company.
4. Barnett, Camille. 1986. *The Creative Manager*. Washington, D.C.: International City Management Association.
5. Cole, Robert E. (Editor). 1995. *The Death and Life of the American Quality Movement*. New York: Oxford University Press.
6. Goldratt, Eliyahu, and Cox, Jeff. 1986. *The Goal*. Croton-on-Hudson, NY: North River Press.
7. International Business Machines. 1984. *Process Control, Capability, and Improvement*. IBM Quality Institute.
8. Nakajima, Seiichi. 1988. *Introduction to TPM*. Portland, Oregon: Productivity Press. (English translation)
9. Ohno, Taiichi. 1988. *Toyota Production System, Beyond Large-Scale Production*. Portland, Oregon: Productivity Press.
10. Schonberger, Richard J. 1982. *Japanese Manufacturing Techniques, Nine Hidden Lessons in Simplicity*. New York: The Free Press.
11. Murphy, Robert; Saxena, Pruneet; and Levinson, William. Semiconductor International. *Use OEE; Don't Let OEE Use You*. September, 1996.
12. Shingo, Shigeo. 1981. *Study of 'TOYOTA' Production System from Industrial Engineering Viewpoint*. Tokyo, Japan: Japan Management Association.
13. Serda, Julian. 1997. *Photolithography Workshop*. Third Annual ATE Conference in Semiconductor Manufacturing, San Jose, CA, July 29, 1997.
14. Studebaker, Don. 1997. *Manufacturing Management Symposium XI*. Sematech, Austin, TX, May 13–15, 1997.
15. Suzaki, Kiyoshi. 1987. *The New Manufacturing Challenge. Techniques for Continuous Improvement*. New York, NY: The Free Press.
16. VanGundy, Arthur B. 1992. *Idea Power*. New York, NY: AMACOM.
17. Van Zant, Peter. 1997. *Microchip Fabrication. A Practical Guide to Semiconductor Processing*, 3rd Edition. New York, NY: McGraw-Hill.
18. Smith, Gerald. 1995. *Statistical Process Control and Quality Improvement*, 2nd Edition. New Jersey: Prentice-Hall.

313

✳ INDEX

5S, 40, 134
6-sigma, 42

Acceptance quality levels (AQLS), 30
Approvals, 161
Assignable cause, 175
Attribute data, 173, 283
   categories, 283
   c-charts, 284
   nonconforming, 283
   nonconformities, 283
   np-charts, 283
   p-charts, 284
   u-chart, 284
Availability, 138, 141
Average, 186

Balanced line, 119
Baldrige, Malcolm, 45
Bathtub curve, 131
   early failures, 131
   stable period, 132
   wearout, 132
Bell-shaped curve, 185
Biased data, 180
Black box, 31

Bottleneck, 86
   gating throughput, 86
Brainstorming, 62
   evaluate ideas, 62
   generate ideas, 62

Calibration procedures, 95
CANDOS, 40, 134
Capable process, 199
Capacity production, 30
   economic order quantity, 30
   economies of scale, 30
Cause-and-effect diagram, 157
Central tendency, 176, 186
Chronic loss
   current accepted condition, 133
   examples, 135
   preventive maintenance, 135
Chronic losses, 132
Clear roles, 61
Collecting good data, 180
Common cause variation, 174, 248
Competitive manufacturing, 10
Competitive market, 9
   commodity market, 9
   three goals, 9

Complementary skills, 51, 61
Constraint, 86
Continual improvement, 38, 127, 165, 209
  reducing limits, 248
Continuous improvement, 107
Control charts
  subgroups, 212
Control limits, 222
  formula constants, 223
  formulas, 222
  sample calculations, 223
Corrective action, 158
  change, 158
Corrective maintenance, 126
Current accepted condition, 133
Cycle time, 87
  series and parallel, 89
  single-piece processing, 90
Cyclic effort, 164

Data, 172, 178
  manufacturing, 178
    biased, 180
    correlation, 179
    good data, 180
    procedures, 179
    sample, 178
    sampling plan, 180
Defect-free, 38
Deming, W. Edwards, 30

Efficiency
  actual effort, 11
  theoretical minimum effort, 11
Eight-step improvement plan, 151
  analyze initial conditions, 152
  collect data, 155
    GIGO, 156
  continually improve, 165
  corrective action, 158
    Pareto chart, 159
    prioritize change, 159
    process platform, 158
    team consensus, 159
  error-proofing, 162
    Poka-Yoke, 162
    visual control, 163

    flowcharts, 153
  implement process control, 163
  obtain approvals, 161
  obvious solutions, 155
  process observation, 153
  process observations
    typical activities, 154
  process platform
    cyclic effort, 164
    initial, 152
    linear effort, 164
    model for continual improvement, 164
  root cause, 156
    brainstorming, 157
    cause-and-effect diagram, 157
    fishbone diagram, 157
      example, 157
    on/off test, 158
  short-term solution, 154
  single-variable change, 161
  team approach, 152
  verify improvement, 162
Elements of manufacturing, 14
  infrastructure, 14
Eliminate waste, 107
  floor layout, 111
  idle time, 109
  nonproductive work, 109
Equipment, 127
  automated, 128
    cluster tools, 128
  manual, 129
  semiautomated, 128
  strategy, 129
Equipment capacity, 78
  WPH, 78
Equipment engineer, 59
Equipment losses, 132
Equipment performance, 130, 135
  availability, 138
  bathtub curve, 131
  chronic loss, 132
  current accepted condition, 133
  equipment measurements, 138
    misleading, 139
  improper maintenance, 137
    examples, 137

improper operation, 137
    examples, 138
incorrect installation, 137
manufacturing problems, 130
MTBF, 138
MTTR, 138
politicized equipment, 140
poor design, 136
sporadic loss, 133
Error-proofing, 162

Fishbone diagram, 157
    example, 157
Floor control, 92
Floor control system, 79
Floor layout
    process-oriented layout, 111
    product-oriented layout, 112
    service-chase, 112
    workbays, 112
Flowcharts, 153
Formal team, 56
    ongoing tasks, 57

Good data, 180
Group technology, 112

Histogram, 181
    cells, 181
    class interval, 181
    example, 182
    lower spec limit, 182
    nominal dimension, 182
    process window, 183
    skew, 182
    upper spec limit, 182
Histograms
    examples, 184
Holistic process view, 99

Idle time, 109
    equipment downtime, 109
    idle humans, 109
    idle parts, 109
Improvement
    corrective maintenance, 126
    high-volume manufacturing, 125

incremental improvement, 126
    quantum leap changes, 127
    strategy, 127
In statistical control, 175
Incapable process, 200
Increased range, 241
Incremental improvement, 126
Individual team skills, 63
    listen and talk, 64
    open-minded, 65
Individuals and moving range, 280
Informal team, 53
Inspection, 80
    inspect in quality, 80
    source inspection, 80
Interchangeability, 7
    mass production, 7
    product drawing, 7
ISO 9000, 43
Ishikawa diagram, 157

Japanese manufacturers, 32
JIT, 5
JIT manufacturing, 40

Kanbans, 40, 93

Large lot sizes, 116
Large sample, 176
Learning curve, 12
Line stop, 85, 242
Linear effort, 164
Lot, 7

Machines, 16
    equipment, 16
Maintenance procedures, 95
Maintenance technician, 58
Making the numbers, 14
Manufacturing, 3
    high-volume, 4
    job shop, 4
    low-volume, 4
Manufacturing data, 178
Manufacturing efficiency, 10
    benchmarking, 11
Manufacturing excesses, 33

Manufacturing improvement, 32, 125
Manufacturing line, 3
Manufacturing methods, 15
Manufacturing productivity, 14
Manufacturing quality, 12
Manufacturing team
    human energy, 52
    members, 52
Market, 22
Market conditions, 23
Market-driven quality, 42
Material, 16
Material requirements planning, 31
Mean, 186
    equation, 186
        statistical function, 186
Measurement errors, 181
Median charts, 282
Methods, 15
    consumables, 15
    procedures, 15
    process specifications, 15
    product specifications, 15
Moving average, 280
MTBF, 138
MTTR, 138

National quality award, 45
Natural variation, 175
Nominal target, 211
    calculation, average, 218
    calculation, range, 220
    sample calculation, 220
Normal curve, 184
    percentages, 191
Normal curve
    probability table, 292
Normal probability distribution, 184

OEE, 141
    actions for low OEE, 301
    availability, 141
        total downtime, 141
    calculating, 143
    focus for improvement, 145
    focus resources, 145
    measurement equation, 141

operating efficiency, 142
    available hours, 142
    equipment utilization, 142
order of collecting data, 143
process considerations, 147
rate efficiency, 142
    actual rate, 142
    theoretical rate, 142
rate of quality, 143
    good output, 143
    total input, 143
reasons for major losses, 143
total time, 141
On/off test, 158
Operating efficiency, 142
Operation-based team, 57
Operations, 3
Operator, 49, 58
Operator cross-training, 95
Optimum condition, 12
Out of control, 231
    corrective action, 242, 245
    error-proof, 243
    gray area, 244
    ignoring action, 245
    ignoring problems, 243
    increased range, 241
    minor shutdowns, 243
    monitoring output, 242
    out of specification, 243
    shifting mean, 240
    stop production, 243
    team action, 235
Out of statistical control, 175
Out-of-spec, 7
Overall equipment effectiveness, 141

Parameters, 17
Pareto chart, 304
    example, 159
Part numbers, 79
Patterns, 237
People, 14
Pitfalls of teams, 67
    lose enthusiasm, 68
    setting an example, 69
    team lacks focus, 70

Population, 171
  sample, 178
  total, 178
Prioritize change, 159
Probability, 175
  event, 195
  mean, 186
  Z statistic, 195
    equation, 195
Process, 3
Process capability, 99, 198, 210, 259
  calculating, 262
  calculating Cp, 264
  calculating Cpk, 268
  compared with control, 261
  C_p, 264
  C_{pk}, 267
  example, Cp, 266
  example, Cpk, 270
  general index formula, 262
  improvement, 270
  interpreting Cp, 265
  interpreting Cpk, 269
  scenarios, 259
  statistical view, 198
Process control, 98, 231
  compared with capability, 261
  everyday example, 232
  statistical view, 197
  three visual rules, 233
Process engineer, 59
Process flow, 92
  pull process, 93
  push process, 93
Process improvement, 245
  change, 250
  common cause, 248
  reducing control limits, 248
  stable to unstable, 248
Process improvement.
  special cause, 247
Process leveling, 119
Process observation, 154
Process platform, 152
  basic model, 152
  change, 250
  cyclic and linear effort, 163

initial, 152
  SPC, 250
Process technician, 58
Process window, 183
Product, 3
Product flow, 5
  continuous flow, 5
  downstream, 5
  just-in-time (JIT), 5
  upstream, 5
Product specifications, 7
Product-based team, 57
Pull process
  kanbans, 93
  line stoppages, 94

Quality, 12
Quality circles, 38
  quality improvement teams, 38
Quality control, 7, 12
Quantum leap changes, 127

Random sample, 181
Range, 188
  calculation, 188
  equation, 188
  interpretation, 188
  nominal target, 211
Rate efficiency, 142
Rate of quality, 143
Raw material, 3
Reducing limits, 248
Repeatability, 8
Research and development, 31
Reverse engineering, 31
Rework, 85
Root cause, 156
Run chart, 280

Sample population, 178
Sampling plan, 172, 178, 180
  control group, 181
  disturb the process, 181
  large sample, 181
  measurement errors, 181
  random sample, 181
Scrap parts, 84

Sematech, 32
Semiconductors, 18
Settings, 17
Setup time, 78
Shewart, Walter, 29
Shifting mean, 240
Shop floor, 3
Sigma, 188. *See* standard deviation
  calculator, 189
  defective parts, 190
  equation, 189
  interpretation, 188
  n vs. *n*–1, 189
  probabilities, 190
Single-variable change, 161
Single-piece processing
  cluster tool, 90
Skew, 182
Source inspection, 80
SPC
  attribute charts, 210
  chart construction, 213
  chart overview, 212
  chart patterns, 237
  common charts, 210
  concept, 210
  continual improvement, 250
  control chart, 295
  control limit interpretation, 211
  control limits, 222, 232
  control vs. spec limits, 211
  feedback control, 232
  goals, 231
  increased range, 241
  individuals and moving range, 280
  lower control limits, 211
  median charts, 282
  moving average, 280
  moving range, 280
  nominal target, 211, 232
  out of control, 212
  process platform model, 250
  random pattern, 232
  reason for, 209
  reducing limits, 248
  run chart, 280
  shifting mean, 240
  subgroups, 212

  table of patterns, 287
  three visual rules, 233
  upper control limits, 211
  variables charts, 210
  visual interpretation, 211
  ways for control, 233
  x-bar & R chart, 210
  x-bar & S chart, 210
Special cause, 247
Special cause variation, 175
Specific task, 56
Specification, 7
  LSL, 7
  nominal dimension, 7
  USL, 7
Specification limits, 7, 182
Specifications, 94
  document release procedures, 95
  manufacturing procedures, 95
  product specifications, 7, 95
Sporadic losses, 133
Stable process, 175
Standard deviation, 188. *See* sigma
Standardization, 6
  discipline, 6
Standards, 95
Statistical process control, 29, 39
  SPC, 29, 39
Statistics, 171
  assignable cause, 175
  attribute data, 173
  average. *See* mean, 186
  bell-shaped curve, 185
  central tendency, 186
  common cause variation, 174
  data, 172
    data point, 178
    sampling plan, 172
  experimentation, 171
  in statistical control, 175
  mean, 186
  median, 187
  mode, 187
  natural variation, 175
  normal curve, 184
    mean, 186
    shape, 186
    variability, 187

normal probability distribution, 184
  curve, 184
out of statistical control, 175
probability, 175
  central tendency, 176
  large sample, 176
  mean, 186
  normal curve, 184
  single coin toss, 175
  three coins, 176
  two coins, 176
  uniform outcome, 176
process control, 171
range, 188
sample size, 186
sigma, 188
special cause variation, 175
stable process, 175
test, 171
unstable process, 175
variability, 187
  sigma, 188
  sigma vs. standard deviation, 188
  standard deviation, 188
variables data, 172
variables vs. attribute data, 173
  examples, 174
variation, 174
Subgroup
  average, 216
  define, 214
  measurement frequency, 214
  range, 216
Subgroups, 212
Supervisor, 49, 58
Support personnel, 60

Takt time, 168
Taylor, Frederick W., 29
Team, 51
Team attributes, 60
  clear decisions, 63
  commitment, 63
  defined goals, 61
  intrateam interaction, 63
  open communication, 60
Team charter, 61
Team consensus, 159

Team leader, 61
  facilitator, 62
Team leadership, 65
  coach, 66
  give credit, 66
  treat people fairly, 66
Teams, 51
  manufacturing teams, 52
Three visual rules, 233
  rule 1, 234
  rule 2, 236
  rule 3, 237
Throughput, 86
  bottleneck, 86
Tolerance, 7
Total customer satisfaction (TCS), 42
Total population, 178
Total preventive maintenance (TPM), 41
Total productive manufacturing, 43
Total quality control (TQC), 38
  categories and concepts, 39
Total quality management
  TQM, 43
Traditional organization, 49
  department, 49
  functions, 49
  informal supervisors, 49
Training, 95
  subject-matter experts, 95

Unstable process, 175

Value, 3
Variability, 187
  range, 188
  standard deviation, 188
Variable, 17
  level, 17
Variables data, 172
Variables vs. attribute, 173
Variation, 174
  common cause, 174
  special cause, 175
Visual control, 163

Wafer fab, 19
  service-chase, 112
  terminology, 296
  workbays, 112

Wafer track system, 112
Waste
    non-value add, 10
Waste elimination
    large lot sizes, 116
    lot size, 114
    off-line manufacturing control, 119
    overproduction, 118
    product defects, 118
    setup times, 115
Whitney, Eli, 29
Work in process, 79
    batch process, 79
    excessive wip, 79
    WIP, 79
Workstation, 3, 78
    work cell, 78
WPH, 78

X-double bar, 218
$\bar{x}$-bar & S, 210, 279
$\bar{x}$-bar & R, 210
    calculate subgroup, 216
    define subgroup, 214
    five steps, 213

grand mean. *See* X-double bar
mean of the averages. *See* X-double bar
mean of the means. *See* X-double bar
nominal targets, 211
R-bar. *See* nominal target
record subgroup, 216
subgroup average, 216
    calculation, 216
subgroup range, 216
    calculation, 216
X-double bar, 218
x-bar & R chart, 210

Yield, 82
    accumulative yield, 84
    field performance, 83
    line yield, 84
    station yield, 83
    total process yield, 84
    wafer fabrication yield, 83
    wafer sort yield, 84

Z statistic, 195
    equation, 195
Zero defects, 37